DEVELOPING SHAPE, SPACE & MEASURES

WITH 9-11 YEAR OLDS

SCHOLASTIC MATHS SKILLS

JON KURTA

Published by Scholastic Ltd
Villiers House
Clarendon Avenue
Leamington Spa
Warwickshire CV32 5PR

Text © 2000 Jon Kurta
© 2000 Scholastic Ltd

1 2 3 4 5 6 7 8 9 0   0 1 2 3 4 5 6 7 8 9

AUTHOR
Jon Kurta

SERIES CONSULTANT
Jon Kurta

EDITOR
Joel Lane

ASSISTANT EDITOR
David Sandford

SERIES DESIGNERS
Anna Oliwa and Sarah Rock

DESIGNER
Paul Roberts

ILLUSTRATIONS
Fred Pipes

COVER ARTWORK
Mark Oliver at Illustration

Designed using Adobe Pagemaker

British Library Cataloguing-in-Publication Data
A catalogue record for this book is available from the British Library.

ISBN 0-439-01775-0

CONTENTS

MEASURES

ABOUT THE AUTHOR

Jon Kurta is a lecturer in mathematics ITT at the University of Surrey, Roehampton. He has written several successful mathematics titles for Scholastic, including *Developing Mental Maths with 7–9 year olds*, *Practising Mental Maths with 8–11 year olds* and *Resource Bank: Handling Data*.

This book is dedicated to my colleagues in the Maths team at Roehampton for so many great ideas.

ACKNOWLEDGEMENT

Thanks to June Guainiere for the 'Follow my shape' activity on page 20.

When asked what constitutes the content of mathematics in the primary school, many people will think solely of arithmetical calculation. This part of the curriculum has been emphasized in the contemporary concern with standards that has led to the development of the National Numeracy Strategy: *Framework for Teaching Mathematics* (March 1999) and the introduction of the 'daily mathematics lesson'. However, it is interesting that in the international studies that provoked such concern (such as the TIMMS study), British children actually performed well in comparison to their peers in spatial topics. This perhaps reflects the strong tradition in British primary schools of work on shape, space and measures – a tradition that embraces the Montessori philosophy, as well as Piagetian notions of 'learning through experience', and which has often involved the innovative use of a range of practical materials (geoboards, geostrips, Poleidoblocs and other resources).

In children's everyday experiences, spatial encounters precede numerical ones. In the pre-school years, children are curious about their environment and keen to become familiar with the look and feel of everything around them. A young child picking up and pulling objects could be estimating their size and weight, as well as exploring the relationship between these attributes. As they learn to speak, children develop appropriate language to help them articulate their developing spatial awareness. Taken a step further, we can see how spatial awareness is as important as numerical awareness in many professions – for example, building, carpentry and engineering.

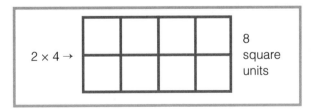

This book provides activities that acknowledge sound practice in the teaching of shape, space and measures, but also aim to embrace contemporary issues. Particular emphasis is given to visualizing aspects of the topics, and to the role of language – both in the sense of accurate use and understanding of 'technical' vocabulary and in the sense of being aware of talk and discussion as an important mode of learning.

SHAPE AND SPACE

DEVELOPING VOCABULARY

SCAA's review of the 1996 Key Stage 2 SATs suggested that teachers needed to place more emphasis on the use of correct mathematical vocabulary for geometric shapes. It is not uncommon to hear children (and adults) referring to a ball as 'circle-shaped' and a box as 'square.' This misconception needs to be addressed, not glossed over. Typical misconceptions that occur in this area are well documented in the OFSTED Research Report*, and many are addressed directly in activities in this book. For example, research shows that children are less likely to recognize and name correctly shapes that are shown obliquely or that are irregular.

WORKING MENTALLY

The difficulties outlined also highlight the importance of talk and discussion. I may know what 'parallel lines' means, and I can close my eyes and picture them, but without dialogue it is unclear whether a group of children have the same mental picture. Spatial activities clearly need to involve a mental element as well as a physical one, and the ability to visualize objects spatially is important in everyday adult life.
It can also be argued that visualizing in spatial activities will be useful for children's development in mental calculation. For example, realizing that the subtraction $82 - 78$ is best done by adding on from 78 may be helped by picturing the position of the numbers on a number line. Other examples include 'knowing', using the dot patterns on a dice to recognize numbers and understanding how multiplication can be modelled with 'arrays' that link directly to area:

$2 \times 4 \rightarrow$					8 square units

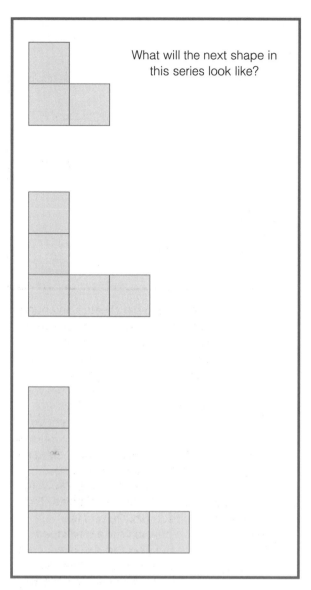

What will the next shape in this series look like?

In the development of algebraic thinking, the use of patterns that combine a spatial element with a numerical element is also common practice. For example, the 'growing L' image is a visual way of representing odd numbers (see figure above).

MEASURES

Work on measures in primary schools is often seen primarily as a context for developing number. This can be seen both in word problems involving use of the four operations (often at the end of a page of 'pure' calculations) and in the use of metric measures to contextualize work on place value and decimals. While not ignoring this aspect, this book puts particular emphasis on conceptual understanding of measures. As with shape and space, there is an emphasis on visualization and discussion. (Further activities on calculations with measure can be found in the companion Scholastic series *Developing Mental Maths* and *Practising Mental Maths*.)

It is important that the teaching of measures should include:

● *Developing a 'feel' for what is being measured and the right words to express it.* This involves firstly linking known words to the appropriate measure (for example, **heavy** and **light** are related to weight, **full** and **empty** to capacity), and secondly having a sense of the relative sizes of different objects – Which is the heavier stone? Which cup holds the most?

● *Being able to make 'common sense' estimates.* In many real-life practical situations involving quantities, we do not need to know an exact measure: for example, when parking my car, putting milk in my coffee or buying a handful of vegetables, I only have an approximate idea of the length of available kerb, the volume of milk or the weight of the vegetables respectively. However, I am confident that I won't hit the car behind, overfill my cup of coffee or buy too many vegetables. It is important to help children build up this 'common sense' of different measures.

● *Systems of units.* Children need to know both the numerical relationships between units (such as 100cm = 1m) and their relative magnitude in practical use. For example, the length of a pencil is best measured in centimetres, but the distance across the playground is best measured in metres. It is important to build up children's images of the relative size of different units.

SOLVING PROBLEMS

While learning about shape, space and measures, children need to be actively engaged in problem solving. This provides a purposeful context for the development of skills and an environment that fosters discussion. Talk needs to be skilfully led in order to make sure the children's attention is focused and misconceptions are addressed. It also provides a vehicle for assessment of the children's understanding.

Additionally, it is through talk that a shared understanding of particular vocabulary can be established. For example, the idea that volume is measured in cm^3 can be built up by talking about how a box can be filled with centimetre cubes, then how the dimensions of the box can be measured and multiplied together. Discussing how these two methods are equivalent helps to establish the cm^3 as a unit of volume.

ABOUT THIS BOOK

This book is split into two main sections: 'Shape and space' and 'Measures'. Each section starts with a double-page glossary (pages 8–9 and 72–73) which highlights relevant vocabulary, particularly where it is likely to be new to this age group. Each glossary can be enlarged for display as an A2 poster (copy each A4 page separately onto A3, then join them back together) or copied at normal size for desktop use. Simpler terms are covered in the two glossaries in *Developing Shape, Space and Measures with 7–9 year olds.*

The two main sections of the book are divided into sub-sections dealing with particular topics. Each sub-section starts with a double-page spread, highlighting:

● **Key ideas.** *What the children need to know overall – the key concepts that need to be considered when planning a unit of work on this topic for the appropriate age group.*

● **Common misconceptions and difficulties.** *Key teaching points, many of which are emphasized in particular activities.*

● **What children should know by the end of Year 5/Primary 6 and Year 6/Primary 7.** *This relates directly to the* NNS: *Framework for Teaching Mathematics, the National Curriculum for England and Wales and the Scottish Guidelines for Mathematics 5–14.*

The activities presented in each sub-section fall into two categories: teacher-directed activities and group problem-solving. If you are following the National Numeracy Strategy, this division is well suited to the structure of the daily maths lesson.

The teacher-directed activities include ideas for introducing particular themes into whole-class discussion, including key questions which can be used for assessment. Some of these activities are short 'warm-ups' to use in the first part of the mathematics lesson; others are suitable for extended discussion in a plenary session following group work on the theme.

The group problem-solving activities are intended to involve the children in group work, and include suggestions for different ability groups (though, as with number activities, this may not always be necessary).

Supporting photocopiable sheets provide either practice activities or resources. The practice activity sheets are designed to consolidate the main concepts. They also stimulate personal reflection or discussion by asking questions as indicated by the following icons:

 think (and be ready to talk) about it!

 discuss with your 'maths talk' partner

 tell your teacher

 write down [to help the children organize their ideas logically].

The use of these photocopiable sheets is intended to be flexible. A class can be split, with half doing a practical activity and the other half working on a practice sheet. The sheets can be used after the completion of practical activities; some may also be useful for homework or individual assessment.

A photocopiable writing frame is provided on page 128. This will help the children to reflect on their learning, and could be particularly useful when they have been engaged in practical or investigative work that has not left much recorded evidence. Activities for which the writing frame is particularly helpful are highlighted in the teachers' notes. The writing frame can be used directly as it is; alternatively, when the children have developed confidence in their writing about maths, they could use it as a basis for a freer account of their work. Asking children to write about maths is a good way to find out how confident they are in their grasp of mathematical vocabulary, and also how able they are to communicate their ideas – two aspects of mathematics which are important themes in this book.

*Askew, M. and Wiliam, D. (1995) *Recent Research in Mathematics Education* London: HMSO.

LINES

Parallel and perpendicular

Parallel lines (never meet):

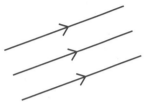

Perpendicular lines (cross at right angles):

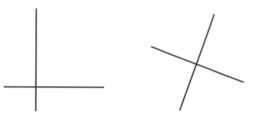

Horizontal and vertical

A **horizontal line** and a **vertical line** form the x-axis and y-axis of a **coordinate diagram** or graph:

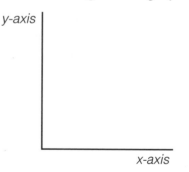

y-axis

x-axis

Coordinate diagrams can be extended to four **quadrants**:

2nd quadrant	*1st quadrant*
3rd quadrant	*4th quadrant*

ANGLES

Angles occur where two lines meet.

They can also be found in the corners of 2-D shapes:

Measuring angles

Angles are measured in **degrees**:
- a full turn is 360°
- a half-turn is 180°
- a quarter-turn is 90° (equivalent to a right angle).

An angle less than a right angle is called **acute** (meaning sharp). An angle more than a right angle is called **obtuse** (meaning blunt).

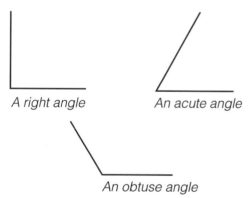

A right angle   *An acute angle*

An obtuse angle

Remember!

- Angles in a triangle add up to 180°. A triangle may have three acute angles, or it may include one obtuse or right angle.
- Angles in a quadrilateral add up to 360°.

SHAPES

2-D shapes (polygons)

There are three kinds of triangle:

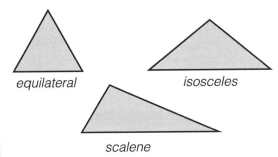

equilateral   isosceles

scalene

A **right-angled** triangle may be either isosceles or scalene:

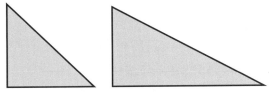

There are many kinds of **quadrilateral**:

square   rectangle   rhombus
parallelogram   kite   chevron
trapezium

Shapes with more than four sides include:
- **pentagon** – five sides
- **hexagon** – six sides
- **heptagon** – seven sides
- **octagon** – eight sides.

3-D shapes (polyhedra)

cube   cuboid   prism
sphere   cylinder   cone
pyramid   tetrahedron octahedron

Circles

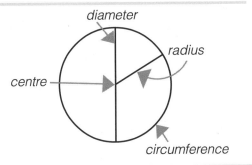

diameter

radius

centre

circumference

TRANSFORMATIONS

Translation

Rotation

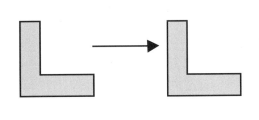

Reflection

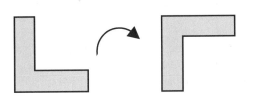

Enlargement

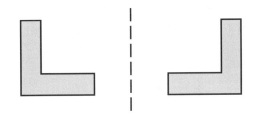

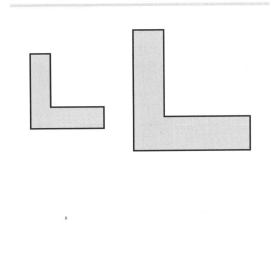

KEY IDEAS

- Classifying and defining a range of 2-D shapes in increasingly formal ways.
- Understanding relationships between 2-D and 3-D shapes (including the use of nets).
- Using a range of practical equipment (including ICT applications) with confidence.
- Developing visualization through games and activities that focus on properties of shapes.
- Exploring shapes creatively – for example, through tessellation and multicultural pattern-making activities.

At this stage, children should be encouraged to use quite formal language in describing the properties of shapes, and to use it in a precise way. Teacher-led discussion is essential to ensure that the appropriate language is modelled for the children. Activities that involve children debating which shapes fit particular criteria are useful for making it clear where their vocabulary needs to be widened. For example, simply saying that a triangle has three sides is not helpful in distinguishing between an isosceles and an equilateral triangle.

Another important teaching tool is to encourage prediction when the children are playing around with spatial ideas. Traditionally, work on shape and space has been characterized as a practical activity – for example, making a cube from a suitable pre-drawn net. Encouraging prediction and challenging children to justify their findings – in this example, predicting which nets will and which will not make a cube, and explaining why or why not – puts reasoning rather than manual dexterity at the centre of the learning.

Encouraging prediction also highlights the importance of visualizing. You need to equip the children (through appropriately structured discussion alongside work with practical materials) to develop a mental appreciation of geometry that complements their practical ability.

Note: work on measuring circles and using the vocabulary 'diameter', 'radius' and 'circumference', although an aspect of shape work, is also related to measures, and is therefore covered in the 'Length and area' section, page 74 (see the activity 'Measuring circles', page 80).

BY THE END OF Y5/P6, MOST CHILDREN SHOULD BE ABLE TO:

- discuss shapes, referring to faces, edges, diagonals and angles
- identify rectangles by their properties
- classify triangles as isosceles, equilateral or scalene
- make shapes with increasing accuracy, including circles
- use vocabulary such as radius, diameter and circumference when talking about circles
- visualize 3-D shapes from 2-D drawings
- identify different nets (for example, to make an open cuboid)
- copy or create tiling patterns with shape templates.

BY THE END OF Y6/P7, MOST CHILDREN SHOULD BE ABLE TO:

- describe and visualize properties of solid shapes, such as parallel or perpendicular lines or faces
- extend their vocabulary to include shapes such as dodecahedron and trapezium, and descriptive terms such as arc
- classify quadrilaterals, using criteria such as parallel sides and equal angles or sides, and know that a diagonal joins opposite corners
- make shapes with increasing accuracy
- use the structural properties of triangles in model-making
- visualize more complex 3-D shapes from 2-D drawings
- identify different nets (for example, to make a closed cuboid).

SOME COMMON MISCONCEPTIONS AND STRATEGIES FOR CORRECTING THEM

RELATIONSHIPS BETWEEN QUADRILATERALS

Children often find it difficult to accept that a square is also a rectangle, or that a rectangle is also a parallelogram. The former shape in each example is a special case of the latter. This difficulty arises if the children have been taught about squares and rectangles first, with an emphasis on what is different about them.

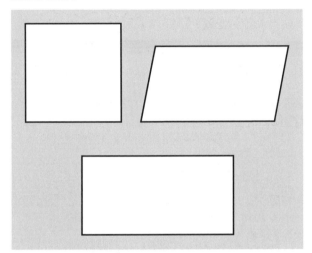

Children need to be encouraged to look for the similarities between shapes as well as the obvious differences – for example, a square and a non-square rectangle (or oblong) both have four right angles; a rectangle and a parallelogram both have two pairs of parallel sides of the same length.

IRREGULAR SHAPES

Children often fail to identify shapes of five or more sides when they are irregular. The popular images of pentagons, hexagons and octagons are invariably of regular shapes.

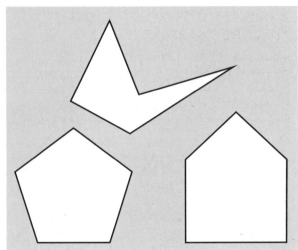

Use examples of irregular shapes, and emphasize that they are 'proper' shapes too. For example, the three shapes above are all pentagons.

DIAGONALS

Children often see all diagonals as lines of symmetry. Again, this tends to occur when regular shapes have been studied first. In particular, many children see the diagonals of rectangles and parallelograms as lines of symmetry:

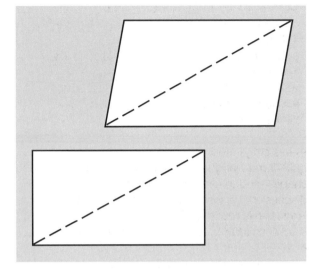

Folding a shape along a diagonal is a good test to see whether the line really is a line of symmetry:

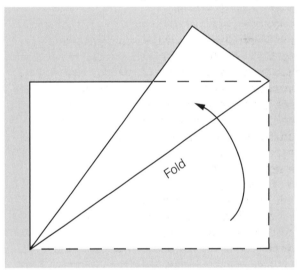

Fold

2-D & 3-D SHAPES

CUT A SHAPE

†† Whole class
⏱ Up to 50 minutes

AIMS
To visualize shapes. To identify key properties of shapes.

WHAT YOU WILL NEED
A4 paper (or thin card), scissors, rulers, pencils.

WHAT TO DO
Before the session, draw a number of rectangles on the board or flip chart. Revise with the children the names and basic characteristics (number of sides and corners) of 2-D shapes. Hold up an A4 sheet of paper or card. Ask them to imagine cutting one or more straight lines in order to create one of the following shapes (select one appropriate to the children's experience): triangle, quadrilateral, pentagon, hexagon, octagon. The straight-line cuts should be from an edge or a corner to another edge or corner. Some examples are shown in the figure below. Depending on the children's previous experiences with this type of visualizing activity, you may need to go through one or two of these examples slowly and carefully to begin with.

Ask the class how they could make each of the shapes. When one child has volunteered an answer, ask the rest of the class to check 'in their heads'. If they are all agreed, draw a line on one of the rectangles on the board to represent the solution. Repeat for the other shapes. Note that many children find it difficult to grasp that a shape such as a hexagon does not have to be regular: any shape with six straight sides will count.

Now ask the children to check the solutions using A4 paper or card. They should use a pencil and ruler to draw the lines across the paper. Encourage careful cutting out with scissors; if card is used, the shapes made can be kept for future reference. The children should experiment to find different solutions for the shapes.

In a plenary session, you might choose to focus on making one particular shape and look at the variety of solutions found.

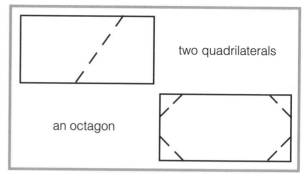

two quadrilaterals

an octagon

DISCUSSION QUESTIONS
● What would happen if I made a cut here... or here... [demonstrating]?
● If you made a cut there, how many sides would the shape have?
● Is this the only way to make that shape?

ASSESSMENT
Can the children visualize and articulate solutions? Can they state the names and properties of the shapes produced?

VARIATION
● Begin with a square or triangular piece of card or paper. How does this change the cuts needed to make the various shapes?

EXTENSIONS
● The children can try to reproduce their shapes on a geoboard. This is particularly useful for helping children to get a feel for irregular versions of shapes such as hexagons or octagons. The geoboard can also be used to model answers to photocopiable page 22.
● The children can use photocopiable page 22 to practise drawing irregular shapes according to various criteria.
● They can use the writing frame on resource page 128 to structure an account of the activity.

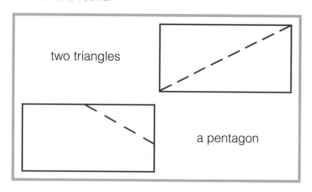

two triangles

a pentagon

NAMING TRIANGLES

†† *Whole class, pairs, individuals.*
🕐 *Up to 50 minutes*

AIM
To reinforce the names and properties of different triangles.

WHAT YOU WILL NEED
Resource page 120, photocopiable page 23, OHP.

WHAT TO DO
Write the following triangle vocabulary on the board or flip chart:

equilateral	acute-angled
isosceles	right-angled
scalene	obtuse-angled

Discuss what each term means. (The activity 'Different angles' on page 59 revises the terms 'acute' and 'obtuse'.)

Use the sheet of triangles (resource page 120) as an OHT, or draw the triangles on the board. Give an oral description such as *Three equal-length sides* or *Three different-length sides with a right angle* and ask the children to identify which triangle you are thinking of. Move towards the use of more formal vocabulary (such as 'Right-angled isosceles triangle'); but continue to mix and match examples and language, since the children will be at different stages of feeling comfortable with the formal triangle names.

Note that the triangles on resource page 120 can be identified by their angles, as well as by their sides – in some cases, both of these are needed (for example, B, D and G are all isosceles triangles). When the children are familiar with this idea, ask for volunteers to describe one of the triangles and ask the rest of the class to suggest which one is being described.

Now the children should work in pairs. The sheet of triangles should **not** be visible. Describe a triangle type, using either formal or less formal language (for example, *An obtuse-angled isosceles triangle* or *A triangle with one angle larger than 90 degrees and two equal-length sides*). The pairs should discuss what the triangle will look like. After a couple of minutes, ask one pair to draw the triangle as they imagined it. Invite other pairs to comment. Repeat several times.

Now distribute copies of page 23. The children should work individually to complete the diagram, then discuss the final problem in pairs. Review the answers in a plenary session.

DISCUSSION QUESTIONS
● *How can you tell the difference between this triangle and that one?*
● *What is the same/different about these two triangles?*
● *Can you explain the difference between an isosceles and an equilateral triangle?*
● *Can you explain the difference betweeen an acute-angled and an obtuse-angled triangle?*
● *What would a triangle with two right angles look like? [It's impossible!]*

ASSESSMENT
Can the children recognize different triangles from descriptions? Can they describe triangles, using formal vocabulary accurately?

VARIATIONS
● Use the quadrilaterals sheet (resource page 121) instead of the triangles sheet, or use both together.
● When describing the triangles, include a particular orientation – for example, *An obtuse-angled scalene triangle with the longest side horizontal.*

EXTENSIONS
● Photocopiable page 23 can be used at another time to assess the children's knowledge of the characteristics of different triangles.
● The first part of this activity can be repeated frequently as a 10-minute game with the whole class. (You and/or the children could give descriptions.)

DEVELOPING SHAPE, SPACE & MEASURES

2-D & 3-D SHAPES

2-D & 3-D SHAPES

NAMING QUADRILATERALS

†† Whole class, pairs
🕐 50 minutes

AIM
To identify the key properties of different quadrilaterals.

WHAT YOU WILL NEED
Resource page 121 (enlarged and displayed, one between two or used as an OHT), a vocabulary list (see below) on the board or flip chart, OHP.

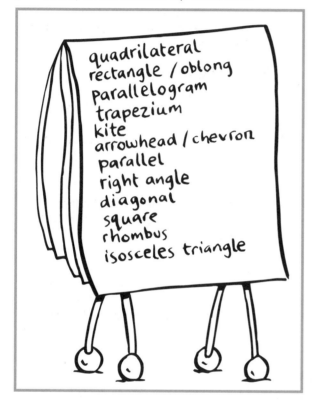

quadrilateral
rectangle / oblong
parallelogram
trapezium
kite
arrowhead / chevron
parallel
right angle
diagonal
square
rhombus
isosceles triangle

WHAT TO DO
Display or give out the quadrilaterals sheet (resource page 121). Ask the children what all the shapes on this sheet have in common. (Four sides, four corners.) Explain that just as 'triangle' is a name for any shape with three sides, so 'quadrilateral' is a name for any shape with four sides. Direct their attention to the vocabulary list. Ask them to work in pairs, giving names to the different quadrilaterals on the sheet. Help them with the more obscure ones.

Once the children are familiar with the names (practise a few times by saying a name and asking children to point to the right quadrilateral), ask them to work in pairs. Each pair should choose a pair of quadrilaterals and identify one thing they have in common (apart from having four sides) and one thing that is different between them. For example,

they may choose A and B and point out that both have four right angles, but B has four equal-length sides while A does not. Characteristics of the shapes that are useful to compare include: number of right angles, lengths of sides, lines of symmetry, lengths of diagonals, parallel lines. Allow five minutes for this, then ask the children to volunteer some examples. Check that the rest of the class agree.

Now ask the children to work with the quadrilaterals sheet, making written notes on the similarities or differences between the shapes. Conclude by giving the children another opportunity to volunteer examples.

DISCUSSION QUESTIONS
● *What do those two shapes have in common? Is there another shape with that characteristic?*
● *What is the difference between these two shapes?*
● *Which shapes have right angles?*
● *Which shapes have parallel lines?*

ASSESSMENT
Can the children identify common characteristics and differences between quadrilaterals? Do they use the names of the quadrilaterals and the words for different properties confidently?

EXTENSIONS
● Children (particularly less confident groups) may benefit from using geostrips or similar resources to make the different quadrilaterals.
● Using an enlarged (A3) copy of page 121, the children can draw lines to link pairs of shapes and write beside each line a common characteristic or difference between the shapes. (They can use different colours for common characteristics and for differences.)

● They can cut up an A3 copy of page 121 and reorder the shapes, mounting them on sugar paper, to highlight common characteristics within groups of shapes or differences between the groups.

● Photocopiable page 24 is a good way to check the children's recognition of different quadrilaterals. Encourage them to visualize each shape before joining the corners. The sheet could be enlarged or copied onto an OHT and used with the whole class. The answers are: a) isosceles trapezium, b) square, c) parallelogram, d) kite, e) trapezium, f) rectangle (or oblong), g) square, h) rectangle (or oblong).

3-D SHAPES CHALLENGE

†† *Whole class, working in pairs*
🕐 *30 minutes*

AIM
To identify the key properties of 3-D shapes.

WHAT YOU WILL NEED
Sets of 3-D shapes (cube, cuboid, sphere, hemisphere, pyramid, tetrahedron, prism, cylinder, cone), at least one shape per pair. You could also use real objects that have these shapes, such as a ball and a can. List the 3-D shapes on the board.

WHAT TO DO
Put all the 3-D shapes on a table or in the middle of the floor, so all the class can see them. Ask the children to discuss with a partner all the different ways that the shapes vary. After a few minutes, ask for suggestions. Encourage the children to go

beyond talking generally about 'colour' and 'size' to using specific mathematical criteria: some have curved surfaces, some have all flat surfaces, some have a mixture; some rise to a point; they have different numbers of sides, edges or corners; some have all right-angled corners and so on. Write all the suggestions onto a flip chart.

Now distribute the 3-D shapes, one per pair. (It doesn't matter if more than one pair have the same shape.) They should keep their shape hidden from the rest of the class, and collaborate to write a description of the shape. Pairs that finish quickly can be given a second shape. When all the pairs have written at least one description, they should read it out; the rest of the class have to identify which shape it is. If another pair have been using the same shape, they should read their description too. Continue until all the pairs have had an opportunity to read a description.

DISCUSSION QUESTIONS
● *What shape has that property?*
● *Are you sure it's that shape – could it be any other?*
● *What part of the description told you what shape it was?*

ASSESSMENT
How accurately do the children describe their shape? Can they identify and name the various 3-D shapes from the descriptions?

VARIATION
Put all the shapes in a large cardboard box. Ask a child to dip in a hand and pull out a shape at random. Without revealing the shape, the child should describe some of its properties until the class are able to identify it.

2-D & 3-D SHAPES

TESSELLATION

†† *Whole class, then individuals*
🕐 *30 minutes*

AIM
To investigate tessellating patterns.

WHAT YOU WILL NEED
Resource page 122, plastic 2-D shapes, squared paper, pencils.

WHAT TO DO
Discuss previous work on tessellation with the children; they should have some previous experience of 'fitting shapes together'. You may want to show them some examples (Escher-type prints), or there may be examples around the school which you can point out (such as hexagonal or rectangular paving patterns). Where children have little experience, it is worth letting them play around with sets of plastic shapes to find out which will tessellate and which will not.

Tell the children that in this activity, they are going to investigate the tessellation of less-regular shapes: the pentominoes. Give out squared paper and copies of page 122. They should choose one pentomino and use it to draw a tessellating pattern on squared paper. The figure below shows an example, using pentomino 8 from the sheet. Note the use of shading (the children can use different colours) to enhance the effect.

In a plenary session, let the children display and discuss their patterns.

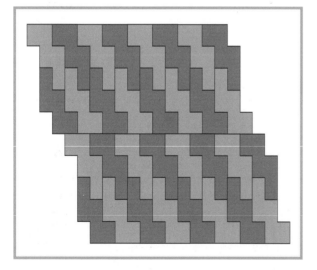

DISCUSSION QUESTIONS
● *What do you remember about tessellation? Which shapes [show different plastic shapes] will tessellate? Which will not?*
● *Which pentominoes produce the most interesting patterns?*

ASSESSMENT
Can the children create a tessellating pattern? Can they explain the characteristics of their pattern?

EXTENSIONS
● The children can try to make tessellating patterns using some of the hexominoes (see resource page 123).
● They can use photocopiable page 25. to practise tessellating shapes, then use squared paper to experiment with other shapes.
● They can investigate tessellating patterns on isometric grid paper (see examples below).

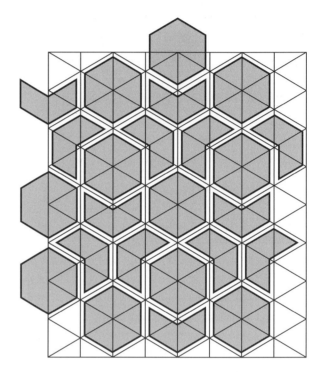

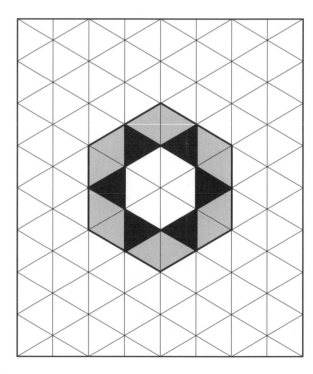

LET'S GO 3-D

†† *Whole class, then groups*
🕐 *50 minutes*

AIMS
To make 3-D shapes. To investigate the number of faces, vertices and edges that each shape has.

WHAT YOU WILL NEED
3-D shapes with plane surfaces only (cubes, cuboids, pyramids and prisms, but not cylinders or cones), Polydron or other interlocking plastic shapes (prepare some 3-D shapes with these before the lesson), straws, Plasticine or play dough, scissors.

WHAT TO DO
Distribute some 3-D shapes. Check that the children are able to identify which parts of the shape are its faces, vertices and edges:

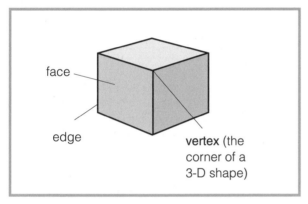

face

edge

vertex (the corner of a 3-D shape)

Now hold up a cube. Ask the children to pair up and see whether they agree how many faces, vertices and edges it has. The number of faces is fairly straightforward, since the children are familiar with a 1–6 dice, but counting (and agreeing on) the numbers of vertices and edges is trickier. When a 3-D shape is turned around in order to count what's at the back, it becomes difficult to keep track of what has already been counted.

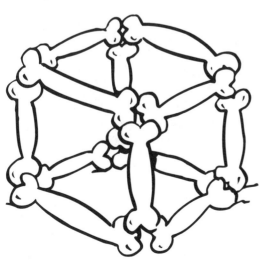

Now demonstrate to the children how to make a skeletal model of the cube. Use cut straw pieces (about 5cm long) for the edges and large pieces of Plasticine or play dough for the vertices (see illustration below). These models should help with the counting of edges and vertices, since these are now distinct objects.

The children should now work in groups, using Plasticine and straws to make skeletal models of a range of 3-D shapes. The 3-D shapes themselves, or models made from interlocking 2-D shapes, can be used as a guide – they are also useful for confirming the number of faces.

Each group should draw up a table (see below), showing the number of faces, vertices and edges on each shape.

SHAPE	FACES	VERTICES	EDGES
Cube	6	8	12
Cuboid			
Tetrahedron			
Square-based pyramid			
Octahedron			
Triangular prism			

In a plenary session, check through each group's answers and resolve any disagreements between groups.

DISCUSSION QUESTIONS
● *How can you make sure you've counted all the faces/vertices/edges?*
● *What is a good way to check the number of each?*

2-D & 3-D SHAPES

2-D & 3-D SHAPES

ASSESSMENT

Do the children recognize the difference between a face, a vertex and an edge on a 3-D shape? Can they work systematically to find the number of faces, vertices and edges on different 3-D shapes? Can they construct 3-D skeletal models efficiently?

EXTENSIONS

● More confident children can examine the table of results and try to see the relationship between the numbers of faces, vertices and edges. This may lead them to discover 'Euler's theorem':
$F + V - 2 = E$. This is true for all 3-D shapes consisting of plane surfaces.
● The children can examine their skeletal models to see which shapes have right angles, and which shapes have parallel edges. Both of these should be clear from the positions of the straws.
● They can use the writing frame on resource page 128 to structure an account of this activity.

OPEN AND SHUT

†† *Whole class, then groups*
🕐 *50 minutes*

AIM

To visualize 3-D shapes from nets.

WHAT YOU WILL NEED

Resource page 122 or 123 (enlarged or copied as an OHT), an A4 copy of the same sheet per pair of children', an open or closed cube (preferably one that can be 'unfolded' to show its net), Polydron or a similar 2-D construction material, scissors.

WHAT TO DO

Depending on the age and experience of the children, you should either use the pentominoes (page 122) to make an open cube (Y5/P6) or the hexominoes (page 123) to make a closed cube (Y6/P7).

Note: Pentominoes and hexominoes are shapes made by joining five and six squares respectively. If time allows, you may want to have the children discover a range of these shapes for themselves. There are 12 unique pentominoes (not counting rotations and reflections), so page 122 contains the complete set. 35 different hexominoes are possible, 11 of which are shown on page 123. These resource sheets are also used for activities on tessellation (page 16), symmetry (page 31) and area and perimeter (page 78).

Show the children an open or closed cube. Ask them to imagine what this shape would look like unfolded. There are several possibilities. Explain that they will be investigating the possibilities in this session.

Ask the children to focus on one particular net: 7 if using the pentominoes, J if using the hexominoes. Ask: *Can everyone see how that net could be used to make a cube shape?* Now ask the children to pair up and discuss which of the nets will make a cube and which will not. Emphasize that they should not cut out the nets, just predict which nets will or will not work. After they have had about five minutes' discussion, ask pairs to compare their predictions. There are usually some nets that everyone agrees about, and others that will be debated.

Now let the children test their predictions. This can be done either by cutting out and folding up the nets from the resource sheet or by making up the nets from Polydron or a similar material. The children should sort the pentominoes or hexominoes into those that work and those that do not; they can be stuck onto a divided sheet of coloured paper for later mounting.

**DEVELOPING SHAPE,
SPACE & MEASURES**

In a plenary session, review the children's findings. The solutions are as follows:
● Pentominoes 2, 3, 4, 5, 7, 8, 10 and 11 will make an open cube; 1, 6, 9 and 12 will not.
● Hexominoes A, D, F, G, J and L will make a closed cube; B, C, E, H and K will not.

DISCUSSION QUESTIONS
● *Can anyone explain how that will make a cube?*
● *Can anyone explain how that will definitely not make a cube?*
● *Which ones were easy to predict? Which were the trickiest?*

ASSESSMENT
Can the children express and justify their predictions, using appropriate mathematical vocabulary?

EXTENSIONS
● If they are using the hexominoes resource sheet, challenge the children to find further arrangements of the six squares that will (or that will not) make a cube.
● When they have identified the hexomino nets on page 123 that will make a cube, challenge the children to place the numbers 1–6 in the squares so that when it is folded up, the cube will have opposite faces that total 7 (as in a real dice).
● Photocopiable page 26 pursues a similar 'visualize, predict and try' theme with possible nets for pyramids (both triangular-based and square-based). The sheet is intended for individual consolidation work, but the nets could be copied onto an OHT or enlarged for whole-class discussion. The shapes will be easier to manipulate if the sheet is enlarged to A3 size.

DICTIONARY AND DATABASE

†† *Whole class, then pairs*
🕐 *30 minutes; subsequent work over 2 weeks*

AIM
To consolidate understanding of the key properties of shapes.

WHAT YOU WILL NEED
Various 2-D and 3-D shapes, copies of the 'Shape and space glossary' (pages 8–9), maths dictionaries, a computer database package. The database will need to be set up with fields representing each of the different aspects arising from the class brainstorm (see below).

WHAT TO DO
Display a range of 2-D and 3-D shapes. Ask the children what aspects pairs of shapes have in common, and in what ways they are different. Brainstorm suggestions. The results should include at least some of the following: 2-D or 3-D, number of sides, number of angles, right angles or not, parallel lines, symmetrical or not, regular or not.

Tell the children that, in pairs, they are going to write a complete description of one of the shapes. The description will need to be carefully drafted and redrafted on the class computer. The final version will form part of a 'wall dictionary': an alphabetically organized display representing all the shapes they have been using. Distribute shapes (one or two per pair). In the initial session, the children should make notes about their shape and attempt a first draft for the dictionary.

Show the class the computer database and explain that when their entry for the wall dictionary is complete, they can use it to enter information (in more condensed form) about their shape on the class database of shapes. You might want to organize a rota for the use of the class computer.

DISCUSSION QUESTIONS
● *What are some of the properties we can use to distinguish the shapes?*
● *What sort of information could we include on the database?*

ASSESSMENT
Can the children suggest suitable criteria for sorting the shapes? Can they complete a concise and accurate description of their shape? Can they use the computer database with understanding?

EXTENSIONS
● Continue to add new fields to the database as the

2-D & 3-D SHAPES

Name	Rectangle
Dimensions	2D
No. of sides or surfaces	4
Number of corners	4
Symmetrical	yes
Parallel lines	yes
Other features	4 right angles, diagonals bisect

children meet new topics, for example symmetry and tessellation.

● The children can use the sort facilities of the database to investigate questions such as: *Which shapes have right angles? Which have symmetry? Which are regular?* They can print lists of shapes under these different categories, and display them alongside the wall dictionary.

● Photocopiable page 27 can be used for individual consolidation of this topic. The children could use either the wall dictionary or the class shapes database to resolve any disagreements. The answers are: 1. False, 2. False, 3. True, 4. True, 5. True, 6. False, 7. False, 8. True.

FOLLOW MY SHAPE

†† *Whole class, working in pairs; groups*
🕐 *50 minutes*

AIM
To practise naming and describing shapes.

WHAT YOU WILL NEED
The gamecards from resource page 124, copied onto card, laminated and cut out (you may prefer to enlarge the sheet to A3 size first), blank cards, pencils.

WHAT TO DO
You may have played similar 'loop' or 'follow me' games using number operations, in which case little instruction will be needed. Shuffle the set of 15 gamecards and give them out, one per pair. All of the cards have to be used each time; so if you have fewer than 30 children, some individuals will have a card each.

Choose a pair to begin. One of them should read out the description on the lower half of the card. Another pair should have that shape's name printed at the top of their card. They should read out the name of the shape; if all agree that it is correct, they should read the description of the next shape from the lower section of their card. The game continues until the class have gone 'full circle', returning to the pair who started the game. Collect in the cards, shuffle them and play again.

If you play a number of times over a week or a

fortnight, you can time the class and monitor their improvement.

In the second part of the lesson, the children should work in groups of five or six. Set them the task of creating their own 'loop' game with five or six cards, focusing on 3-D shapes. They should work together on a set of descriptions. They will need to check that each description has only one possible answer – for example, 'I have all right angles' is not accurate enough because it could be either a cube or a cuboid. When they have produced five or six descriptions, they should make up the set of cards. Groups can swap sets and try out each other's games.

During a plenary session, discuss with the children how the different clues help them to work out which description of a shape matches which shape name.

DISCUSSION QUESTIONS
● *How did you know that that was your shape?*
● *Are you sure that description could only be a cube?*
● *Which descriptions are easy to get? Which are more tricky?*

ASSESSMENT
Can the children identify the shapes? Can they write appropriate descriptions of shapes?

VARIATIONS
Use the gamecards for whole-class quizzes:
● Read each description and invite suggestions as to what the shape is.
● The children try to identify each named shape by asking you questions to which you will only answer *Yes* or *No*.

KEEP OFF THE GRASS

DRAW THE SHAPE

■ Using a pencil and a ruler, join the dots to draw a shape that matches each description. You might want to model the shapes on a geoboard first.

A pentagon with a right angle.

A symmetrical octagon.

A concave hexagon.

A heptagon.

A shape with more than eight sides.

A hexagon with two right angles.

 Use dotty paper to set some more challenges like this for your friend to solve.

SEE 'CUT A SHAPE', PAGE 12.

**DEVELOPING SHAPE,
SPACE & MEASURES**

TRIANGLE SORT

■ Copy each of these triangles into the correct cell in the sorting diagram below:

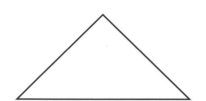
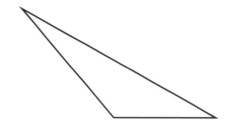

■ Now draw suitable triangles in as many of the other cells as you can.

Length of sides

	3 equal	2 equal	All different
Acute < 90°			
Right = 90°			
Obtuse > 90°			

Largest angle size

 Explain to your friend why it is impossible to have a right-angled equilateral triangle.

SEE 'NAMING TRIANGLES', PAGE 13.

DEVELOPING SHAPE,
SPACE & MEASURES

NAME   DATE

JOIN THE DIAGONALS

■ Using a pencil and a ruler, join the ends of each pair of crossed lines to make a quadrilateral. Try to predict first which quadrilateral it will be.

a)

b)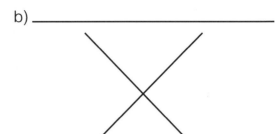

c)

d)

e)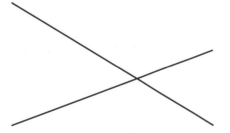

f)

g)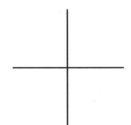

h)

 Which ones were easy to predict? Does your friend agree?

SEE 'NAMING QUADRILATERALS', PAGE 14.

MAKE IT TESSELLATE

■ Complete these tessellating patterns:

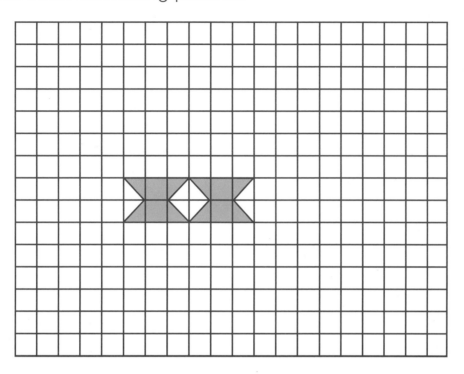

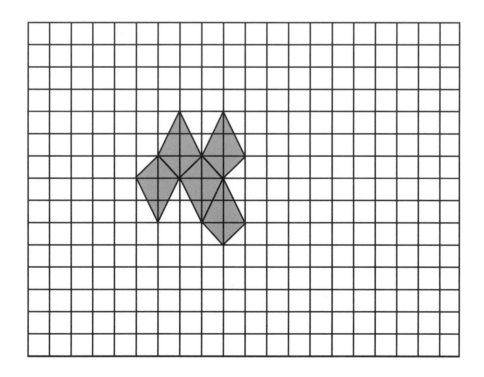

 Now design your own pattern, using squared paper.

SEE 'TESSELLATION', PAGE 16.

**DEVELOPING SHAPE,
SPACE & MEASURES**

NAME   DATE

SEEING PYRAMIDS

■ Look at the nets below and predict which ones will form a pyramid.
It could have either a triangular base or a square base.
■ Discuss your predictions with a friend. Do you agree about them all?
■ Now cut out the shapes and check whether you were right.

Net	Prediction: definitely/definitely not/maybe	Yes/No
A		
B		
C		
D		
E		
F		

 Which ones surprised you? Why?

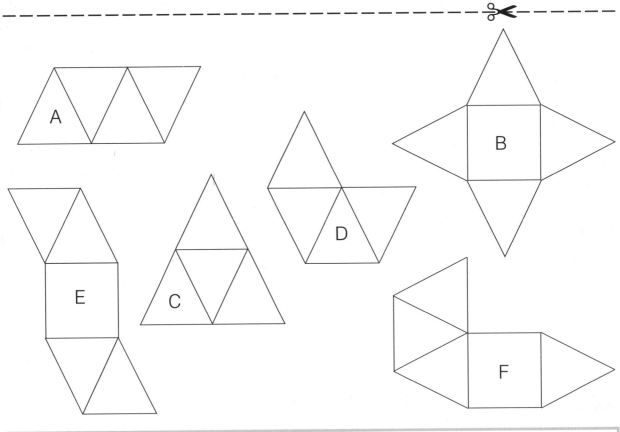

SEE 'OPEN AND SHUT', PAGE 18.

2-D & 3-D SHAPES

WHO'S RIGHT?

■ Four of these children are right, four are wrong. Can you work out which is which? Check any words you are not sure of in a dictionary.

1. All pentagons have six sides.

2. All pyramids have a square base.

3. All rectangles have four right angles.

4. All triangles are convex.

5. A parallelogram has two pairs of equal-length sides.

6. A cylinder has no edges.

7. A kite has parallel lines.

8. A triangle cannot have two right angles.

 Does your friend agree?

 Make up another quiz like this for your friend to try.

SEE 'DICTIONARY AND DATABASE', PAGE 19.

KEY IDEAS

- Recognizing the differences between various transformations, and starting to know their names.
- Rotating and reflecting shapes using coordinates in four quadrants.
- Reflecting shapes in diagonal as well as horizontal and vertical lines.
- Rotating and reflecting 3-D objects.
- Recognizing and completing symmetrical patterns.

Work on this topic at this stage needs to focus on transformations of increasing complexity. Children need to be confident in the use of the associated vocabulary (see the 'Shape and Space' glossary on pages 8–9). Most of the activities will involve children transforming shapes or objects in different ways; but it is equally important for developing the correct use of language to show children the results of some transformations and asking them to describe the effects. Children should also have opportunities to predict the effects of different transformations. The picture below shows the effects of reflecting and rotating an image – it is useful to consider what is similar and what is different between the two transformations.

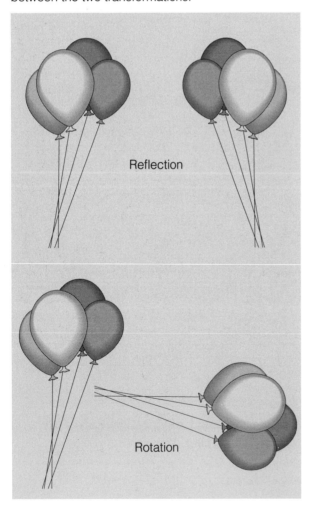

Reflection

Rotation

Work on transformations can usefully be linked to the use of coordinate grids. Children can be asked to alter one or other of the coordinate points of a given shape – for example, in the diagram below, adding 3 to each horizontal coordinate will slide the shape to the left, while adding 3 to each vertical coordinate will slide it upwards.

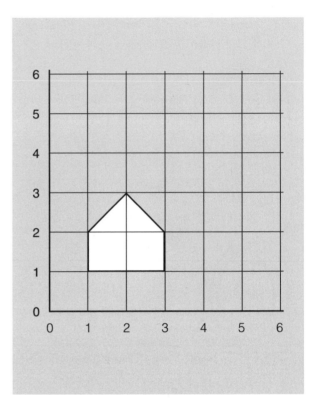

The use of coordinate grids also allows for a structured exploration of the effects of moving the line of reflection. For example, consider the different effects of reflecting the shape in lines A, B and C in this diagram:

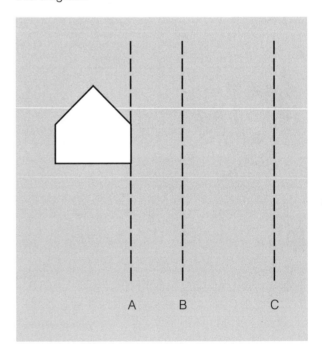

A   B   C

BY THE END OF Y5/P6, MOST CHILDREN SHOULD BE ABLE TO:

● recognize reflective symmetry in regular polygons, including ones with several lines of symmetry
● complete symmetrical patterns with two lines of symmetry at right angles
● reflect shapes in a mirror line parallel to one side
● translate shapes using a coordinate grid.

BY THE END OF Y6/P7, MOST CHILDREN SHOULD BE ABLE TO:

● recognize where a shape will be after reflection (from any starting position)
● recognize where a shape will be after two translations
● recognize where a shape will be after a rotation through 90° about one of its vertices
● make and investigate a general statement about translating or reflecting familiar shapes
● use coordinate grids with four quadrants.

COMMON MISCONCEPTIONS AND STRATEGIES FOR CORRECTING THEM

REFLECT OR ROTATE?

Children often confuse reflective and rotational symmetry. Remind them to think of **mirrors** for reflection and **turning things** for rotation.

MOVING THE MIRROR LINE (1)

Children may have difficulty with identifying a line of reflection where this differs from the line of symmetry of a single shape. For example, if the square on the left is reflected in the mirror, it and its reflection should be equidistant from the mirror:

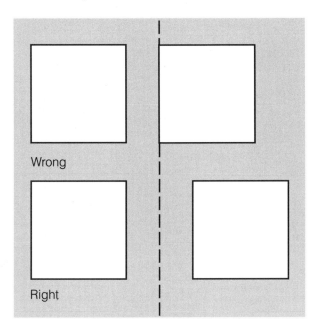

MOVING THE MIRROR LINE (2)

Difficulties with identifying the line of reflection are exacerbated by varying the angle of reflection; some children will just ignore the angle of the mirror:

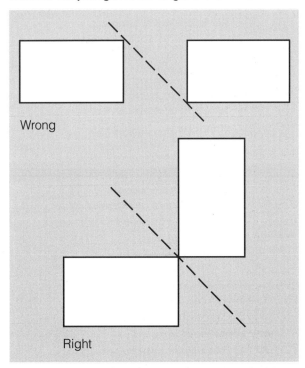

In both of the above cases, it is useful to pre-empt any difficulties by drawing attention to these potential errors when introducing the work. Research also suggests that using squared paper to draw out the results of rotations and reflections helps children to visualize them. It will also be helpful to check with a mirror when visualizing of reflection is difficult.

TRANSFORMATIONS & SYMMETRY

TRANSFORM THE TRIANGLE

👥 *Whole class, working in pairs*
🕐 *50 minutes*

AIMS
To revise the concepts of rotation and reflection. To explore various transformations of a single shape.

WHAT YOU WILL NEED
3 × 3 geoboards and elastic bands, resource page 125. A further copy of page 125 should be enlarged and displayed on the board or flipchart (or copied onto an OHT).

WHAT TO DO
Ask the children how many different triangles they could make on a 3 × 3 geoboard. Reflections, rotations and translations are not allowed. Pairs should use a geoboard, elastic bands and a copy of page 125 to make and record their triangles. Spend only a few minutes on this, as it is not necessary for each pair to find all the possible triangles. Then ask one pair to draw up one of their triangles on the enlarged or OHT grid. Each subsequent pair should attempt to draw a triangle that has not been shown yet. Fairly soon, it should become apparent that for most of the triangles, several different rotations, reflections and translations are possible.

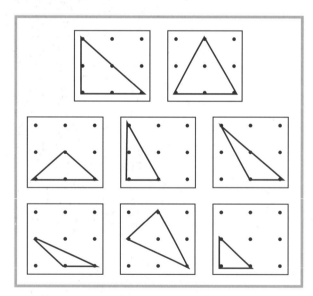

When it has been agreed that there are eight different triangles (see figure above), ask the children to look at them one at a time and to find as many different positions on the grid for that one triangle as possible. You might want to go through an example of this with the whole class first; all the solutions for one triangle are shown below. The children might not record their solutions as

systematically as this to begin with, but they should be encouraged to try (using a fresh copy of page 125). With this kind of systematic recording, it is easier to see how the different transformations of the shape relate to each other (which is a rotation, reflection or translation of another). Pursue this latter idea in a plenary session.

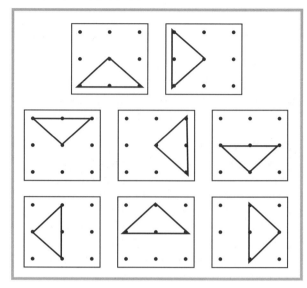

DISCUSSION QUESTIONS
● *Are you sure that this triangle is different from that one?*
● *Which triangle do you think will have the most different positions?*
● *How can you change this position of the triangle to that one?*
● *Can anyone give me an example of a rotation/ reflection/translation?*

ASSESSMENT
Can the children find a range of reflections, rotations and translations for each triangle? Can they describe the relationships between the different placings of a single triangle, using appropriate mathematical language?

EXTENSIONS
● The children can investigate with quadrilaterals. Sixteen unique quadrilaterals are possible on a 3 × 3 geoboard.
● They can use a 4 × 4 or 5 × 5 geoboard, recording on regular dotty paper. This is best attempted when the children are sure of a systematic way of recording.

SHAPE SYMMETRY

†† *Whole class, then groups*
🕐 *Up to 50 minutes*

AIM
To explore the order of symmetry of plane shapes.

WHAT YOU WILL NEED
Resource pages 120 and 121, mirrors or scissors. For whole-class discussion, an OHT or enlarged version of each resource sheet would be useful.

WHAT TO DO
Ask the children to look at one of the resource sheets and to identify any shape that has a line of symmetry. For the purposes of this activity, this can be explained as meaning that, were a mirror placed along the line of symmetry, the reflection in the mirror would be identical in every way (size and orientation) to the original half. Another way to demonstrate this is to fold a paper cut-out of a shape along the predicted line of symmetry. If the shape is symmetrical along that line, it will fold exactly across itself. The latter method is particularly useful for resolving disagreements – for example, many children assume that any rectangle has four lines of symmetry (as a square does), but in fact it only has two (the diagonals are not lines of symmetry). Use one of the triangles or quadrilaterals to demonstrate either of these methods for checking reflective symmetry.

Discuss how many lines of symmetry each shape on the resource page has. Draw up a table of the children's predictions for each shape, and encourage them to explain how they are sure (that is, how they can visualize the lines of symmetry).

Distribute copies of both resource sheets, mirrors and scissors. (Alternatively, you may prefer to have the children consider only the sheet that has not been discussed.) Ask the children to make and investigate predictions, using whatever method they feel comfortable with. They can work individually; if they are working as a group, challenge them to come to a collective agreement about each shape.

Review the children's findings in a whole-class plenary session. The shapes on pages 120 and 121 have the following numbers of lines of symmetry:
Triangles: A) 3, B) 1, C) 0, D) 1, E) 0, F) 0, G) 1.
Quadrilaterals: A) 2, B) 4, C) 0, D) 2, E) 0, F) 1, G) 1, H) 1.

DISCUSSION QUESTIONS
● *Can you explain to your neighbour where the line of symmetry is in that shape?*
● *Which quadrilateral has most lines of symmetry?*

Which triangle? Why do you think that is?
● *What do you notice about all the triangles with one line of symmetry?*
● *Is it easier to recognize symmetry in triangles or in quadrilaterals?*

ASSESSMENT
Can the children predict where the lines of symmetry are for a range of triangles and quadrilaterals? Are they able to check their predictions with a mirror or by folding paper? Are they able to make general statements about their findings, such as 'All isosceles triangles have one line of symmetry'?

EXTENSIONS
● The children can use the writing frame on resource page 128 to structure an account of this activity.
● Photocopiable page 36 can be used to explore line symmetry in a range of pentagons, hexagons and octagons. The shapes have the following numbers of lines of symmetry: A) 5, B) 6, C) 2, D) 8, E) 1, F) 2. The children should notice that for regular polygons, the number of lines of symmetry is the same as the number of sides – thus a square and an equilateral triangle have four and three lines of symmetry respectively.
● Page 40 is a useful means of assessing children's understanding of the ideas covered in this activity and in 'Moving the mirror' (page 34): the children have to spot and correct errors made by a fictitious child.

**DEVELOPING SHAPE,
SPACE & MEASURES**

TRANSFORMATIONS & SYMMETRY

TRANSFORMATIONS & SYMMETRY

TURN IT ROUND

†† *Whole class, then individuals*
🕐 *50 minutes*

AIM
To explore rotations of different shapes.

WHAT YOU WILL NEED
Resource page 122, squared paper, pencils, rulers, tracing paper. Make an OHT of squared paper, or draw a grid of squares on the board or flipchart.

WHAT TO DO
Demonstrate the activity using an OHT of squared paper. Draw two bold lines intersecting at right angles in the centre of the paper to create four quadrants. Choose one of the pentominoes from page 122 and draw it on the grid in the top right-hand quadrant (see Figure 1 below). The shape should be drawn as close to the centre as possible, and lined up with the grid.

Ask children to describe the position of the shape after you have rotated it 90° to the right around the centre point of the diagram. Follow their directions to make the rotation and draw the pentomino on the grid in its new position (which should be as shown in Figure 2). Now consider two further rotations of 90°, again encouraging children to describe the results before drawing on the new position of the pentomino; Figure 3 shows what the result should be. Note that one further, final turn of 90° will turn the pentomino back to its original starting position. The children should discuss this fact, and use it to check their work.

Some children (and adults) find this a very difficult task, so you may want to teach the class the 'trace and turn' method. Trace the original pentomino, then place a sharp pencil point at the centre of the grid and carefully turn the tracing paper 90° to the right. The pentomino will now be in the new position – lift off the tracing paper and sketch it. Repeat for two more turns of 90°. This may take a little practice, but with care can be used to produce effective results.

In the next part of the lesson, the children should practise this individually, working on squared paper and using any of the pentominoes from page 122. It is not necessary for every child to use every pentomino, and you may want to direct less confident children to the simpler shapes (such as pentominoes 1, 2, 5 and 10) on the sheet. Ask the children to make a neat copy of a rotation they would particularly like to show the class.

During a plenary session, share the children's best examples. They can also be used to make an attractive display.

DISCUSSION QUESTIONS
● *What will the shape look like when it is turned 90°? Can you explain why?*
● *What will happen if we make a fourth turn of 90°?*
● *Which pentominoes will be difficult to draw when rotated? Which will be more straightforward?*

ASSESSMENT
Can the children describe the positions of shapes after a rotation of 90°? Can they do so after further 90° rotations? Can they make accurate drawings to show successive rotations?

VARIATIONS
● The children can rotate a range of plastic 2-D shapes (triangles, quadrilaterals and so on) in the same way, drawing around them on blank paper.
● They can rotate hexominoes (see resource page 123) – some of these may be very tricky, but more confident children will enjoy the challenge.

EXTENSION
● Photocopiable page 37 offers further individual practice in rotating shapes.

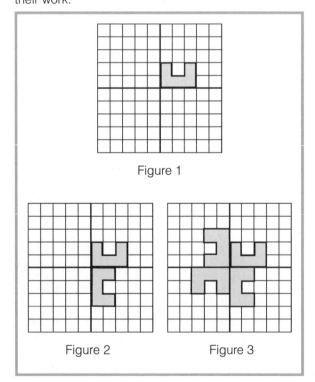

Figure 1

Figure 2   Figure 3

MOVING THE MIRROR

†† *Whole class, working in pairs*
🕐 *Up to 50 minutes*

AIM
To explore mirror reflections.

WHAT YOU WILL NEED
A variety of 2-D plastic shapes, plane mirrors, paper, pencils, rulers, a large mirror for demonstration.

WHAT TO DO
Explain to the children that in this session, they will be predicting the effects of reflecting shapes. Ask them to consider what factors to do with the mirror might influence the results. Depending on their previous experience, the ideas may come from the children or need to be suggested by you. The factors that are useful to consider are:
● the distance between the shape and the mirror
● the angle at which the mirror is placed
● the angle at which the shape is placed.
The different possibilities for shape and mirror placement are shown in the diagram below, which could be drawn on the board as a starting point for discussion. (The first diagram represents the basic arrangement; the other three are all variations.) The diagrams could be repeated with a different starting shape, such as a rectangle or a triangle.

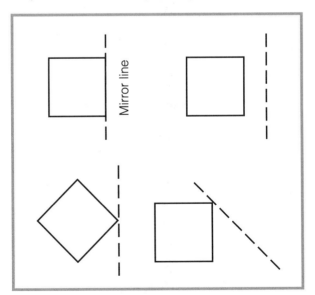

The children should work in pairs to investigate the effect of using different shapes and mirror positions. Less confident children could use a limited range of shapes, with a fixed angle between the shape and the mirror; more confident children could be given a wider choice of shapes and encouraged to vary the angle. It is important to encourage them to predict

(by visualizing) before actually making the reflections. After discussing their prediction with their partner, they should sketch what they think the reflection will look like. They can then take turns to hold the mirror in place while their partner makes an accurate sketch of the results. It is easier to sketch the predictions and results on squared paper or centimetre squared dotty paper than on plain paper; but you might want to give the children a choice, or allow them to discover the value of using squared paper for themselves. Encourage them to be accurate in their representation of the actual results.

During a plenary session, focus on any difficulties the children have experienced with any particular shape. The examples with obliquely placed mirrors are difficult for many children, and may warrant a second session dedicated to exploring this effect.

DISCUSSION QUESTIONS
● *What will this shape look like in the mirror?*
● *What if you turned the shape?*
● *What if you changed the angle of the mirror?*

ASSESSMENT
Can the children predict the likely reflections of a range of shapes? Can they predict, observe and explain the effects of varying the shape's orientation or the placement of the mirror?

EXTENSIONS
● The children can write an account of this activity, using the writing frame on resource page 128.
● They can explore mirror reflections of pentominoes and hexominoes (see resource pages 122 and 123.)
● They can explore mirror reflections of 3-D shapes (making predictions and describing the results to each other if sketching is too difficult).
● Photocopiable pages 38 and 39 can be used for further individual practice in predicting reflections. The examples on page 38 are more straightforward (vertical mirror lines only).
● Photocopiable page 40 can be used to assess the children's understanding of this activity and 'Shape symmetry' (page 31).

TRANSFORMATIONS & SYMMETRY

TRANSFORMATIONS & SYMMETRY

SYMMETRY OF 3-D SHAPES

†† *Whole class, working in pairs*
🕐 *50 minutes*

AIMS
To explore symmetry in 3-D shapes. To construct symmetrical 3-D models.

WHAT YOU WILL NEED
A collection of 3-D shapes, boxes and other objects, Multilink cubes or similar construction materials.

WHAT TO DO
Symmetrical patterns occur frequently in buildings and other constructions; there may well be examples in your local environment that would be useful starting points for discussion. Sometimes the symmetrical pattern is purely decorative; but in many cases, it is an important design feature from the point of view of strength or stability.

Ask the children to consider the 3-D shapes and objects that you have collected – can they identify symmetrical aspects of these? After some discussion, explain that they will be using Multilink cubes (or similar materials) to make a symmetrical 3-D design. Emphasize the mathematical accuracy required: not a cube must be out of place! Say that you will be looking for the most inventive and attractive design.

Working in pairs, the children should create a symmetrical design using 20 cubes (this number can be varied for more or less confident pairs). The children could be given a choice of using one or more colours of cube in their design.

In a plenary session, invite children to show their designs and explain the symmetrical aspects. It might be useful to have pairs check each other's designs before this.

DISCUSSION QUESTIONS
● *How can you tell when a 3-D shape is symmetrical?*
● *How can you check that your design is exactly symmetrical?*
● *Which designs are the most attractive?*

ASSESSMENT
Can the children identify symmetry in 3-D objects? Can they relate symmetry to good design? Can they design and make symmetrical 3-D constructions, and identify the symmetrical aspects?

EXTENSIONS
● The children can sketch their constructions, indicating the line(s) of symmetry.
● They can design symmetrical 3-D objects using other construction materials, such as Polydron or Connect-O-Straws.
● Links could be made to design and technology tasks in which a symmetrical design would be useful – for example, a model bridge, a pencil case or a lunchbox.

DESIGN A LOGO

†† *Whole class, then groups*
🕐 *50 minutes*

AIM
To complete a symmetrical pattern with two lines of symmetry at right angles.

WHAT YOU WILL NEED
Squared paper, coloured pencils, mirrors, some commercial products with logos.

WHAT TO DO
Draw a grid on the board or flip chart (alternatively, make an OHT of squared paper). Draw a bold horizontal line and a bold vertical line to represent two mirrors. Divide the area into four quadrants and create a pattern in the top left-hand corner (see Figure 1 below). Ask for a volunteer to come and draw the reflection of the shape in the vertical axis (as in Figure 2). Now that the top half of the grid is complete, ask for another volunteer to draw the reflection in the horizontal axis (as in Figure 3).

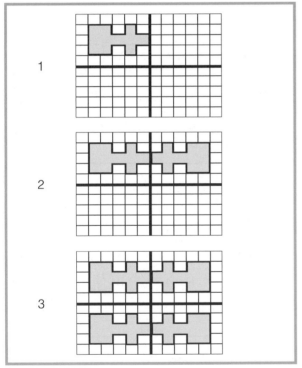

Now ask the children to experiment with their own symmetrical designs, using squared paper. Tell them that they can use their favourite pattern as their own personal 'logo'. Show some examples of company logos on food and drink products. The children can embed their initials in the logo (reflected along with the pattern) to personalize the design further, and use different colours to enhance the effects. The best logos can be displayed (in enlarged form, if the school has a colour photocopier).

DISCUSSION QUESTIONS
● *What happens if you reflect the shape first horizontally, then vertically? [There are two shapes, then four.]*
● *Are you sure that your design has two lines of symmetry?*

ASSESSMENT
Can the children use a range of transformations in a creative way?

EXTENSIONS
● The children can use a computer drawing or painting program to create a symmetrical design (most have a facility for reflecting shapes). Remind them to copy the shape before transforming it, so that they can compare the transformed version with the original.
● Photocopiable page 40 provides further individual practice in making designs by reflecting a shape in the vertical and horizontal axis.

**DEVELOPING SHAPE,
SPACE & MEASURES**

TRANSFORMATIONS & SYMMETRY

NAME   DATE

DRAW THE LINES

■ Draw **all** the lines of symmetry on each of these shapes. You might want to use a mirror to help you.

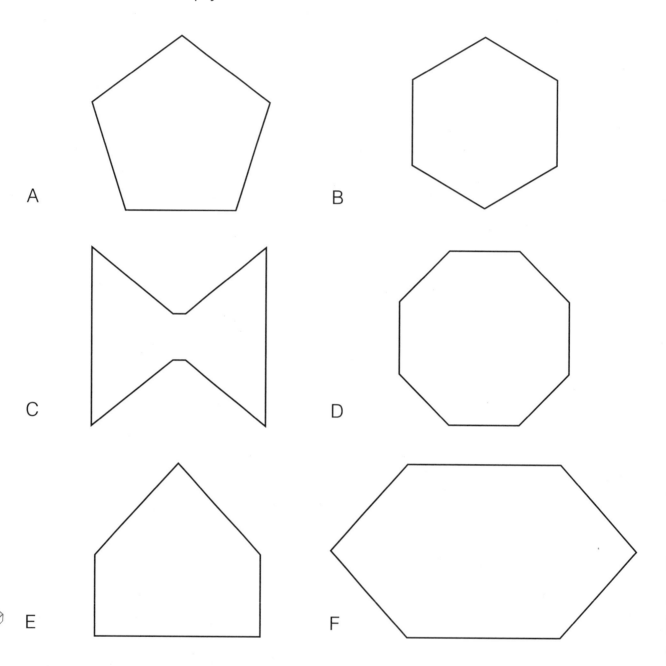

A

B

C

D

E

F

 On the back of the sheet, draw some more pentagons, hexagons and octagons. Can you draw one of each with no lines of symmetry, then one line, two lines and three lines?

 Discuss your answers with your teacher.

SEE 'SHAPE SYMMETRY', PAGE 31.

TRANSFORMATIONS & SYMMETRY

ROTATE A SHAPE

■ Each of these shapes should be rotated 90° three times around the centre of the grid. Draw each shape in its new position in each quadrant, following the rotations. You may find it helpful to use tracing paper.

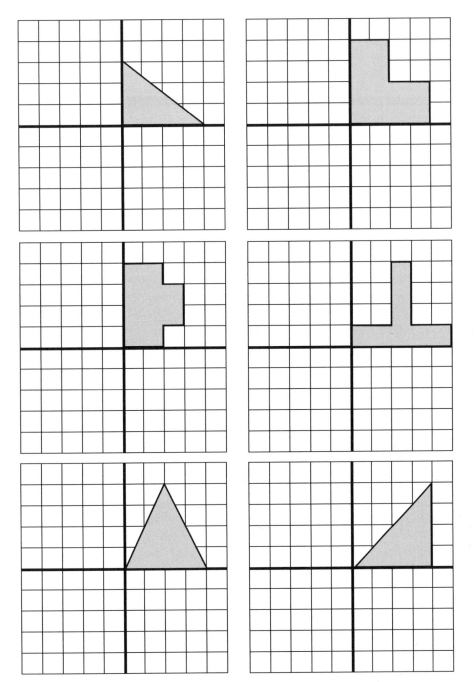

 On the back of the sheet, write some instructions for a friend who has been away from school, explaining how you carried out the rotations.

SEE 'TURN IT ROUND', PAGE 32.

**DEVELOPING SHAPE,
SPACE & MEASURES**

NAME

DATE

REFLECT A SHAPE (1)

■ Draw the reflection of each shape in the mirror line next to it.

 Which ones did you find hardest? Which were easiest?

 Make up some more puzzles like these – four easy ones and four that are difficult.

SEE 'MOVING THE MIRROR', PAGE 33.

TRANSFORMATIONS & SYMMETRY

REFLECT A SHAPE (2)

■ Draw the reflection of each shape in the mirror line next to it.

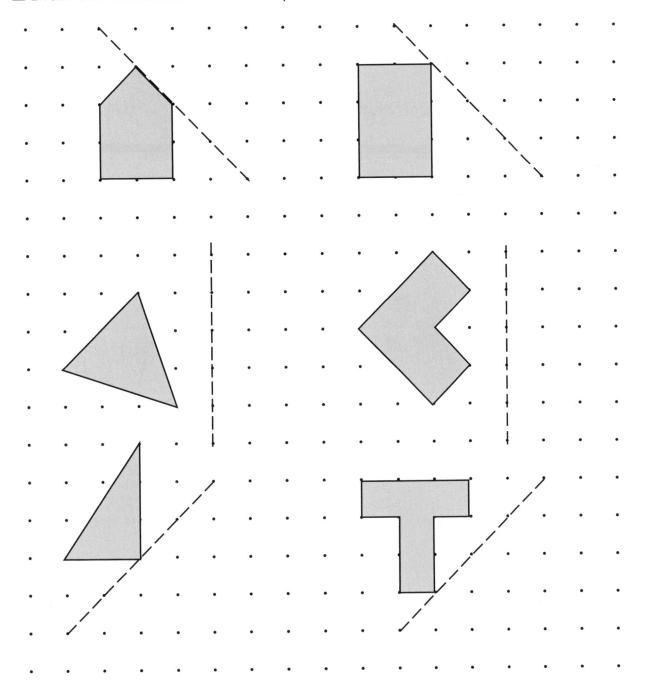

 Choose two or three of the examples and write down some instructions to help a friend who is stuck with these.

SEE 'MOVING THE MIRROR', PAGE 33.

**DEVELOPING SHAPE,
SPACE & MEASURES**

TRANSFORMATIONS & SYMMETRY

TRANSFORMATIONS & SYMMETRY

XARNI'S NOTEBOOK

Xarni the alien is confused about symmetry and reflections. He has tried to fill in this problem sheet, but without much success.

■ On another sheet of paper, draw pictures to show what he should have done.

1. Draw the lines of symmetry on these shapes.

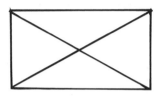

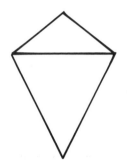
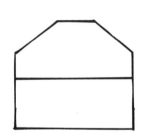

2. Draw the reflection of each shape in the mirror line.

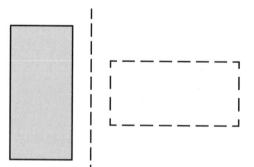
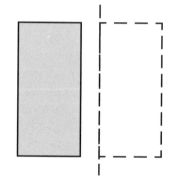
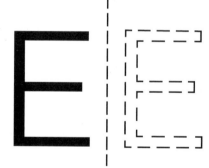
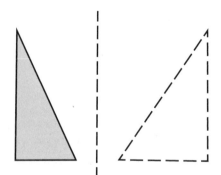

SEE 'SHAPE SYMMETRY' AND 'MOVING THE MIRROR' PAGES 31 AND 33.

**DEVELOPING SHAPE,
SPACE & MEASURES**

FOUR QUADRANTS

■ In each grid, reflect the pattern in the top left-hand corner in the vertical axis, then in the horizontal axis.

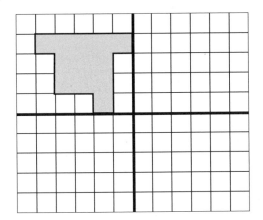 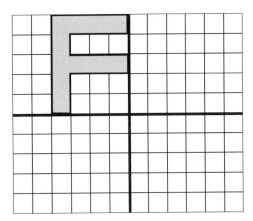

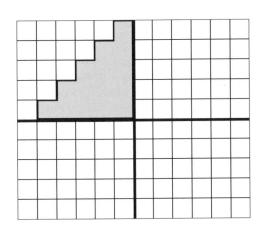 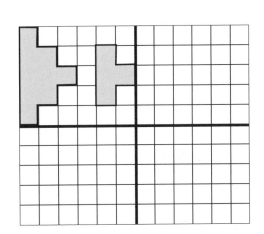

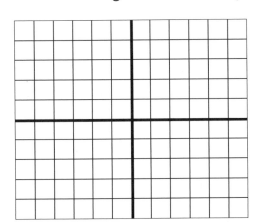 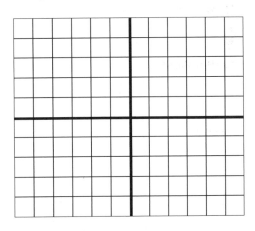

■ Use these grids to create your own symmetrical designs:

 On the back of the sheet, write some instructions for a friend who has been away from school, explaining how to create these patterns.

SEE 'DESIGN A LOGO', PAGE 35.

TRANSFORMATIONS & SYMMETRY

KEY IDEAS

- Using appropriate language to describe position and movement.
- Knowing the compass points (eight directions: N, NE, E, SE, S, SW, W, NW).
- Using coordinates in all four quadrants.
- Knowing the names and properties of parallel and perpendicular lines.
- Using a computer environment (Logo) to give instructions.

At this stage, the children's understanding of position and direction needs both to be reinforced in practical situations and to be challenged in more formal ways. For example, they should be able to plot points confidently on a coordinate grid, and also be developing the ability to visualize and describe positions and locations on the grid in a relative way: 'The position (3, 4) on a coordinate diagram is above and to the right of the position (2, 3).'

More formal language use should also be evident – for example, in discussion about perpendicular and parallel lines, and in the use of 'x-axis' and 'y-axis' to describe the horizontal and vertical position respectively on a coordinate diagram.

Work on these aspects of shape and space can usefully be linked to other areas: work on coordinate grids can be used alongside work on transformations and symmetry (see page 28), and work on parallel and perpendicular lines should reinforce work on the properties of shapes (see the section on '2-D and 3-D shapes', page 10). However, it is important to be aware of the particular difficulties that can arise with each topic.

The children should also be using the eight compass points confidently. This, together with work on coordinates, angles and bearings, should be linked to map work in geography. For example, when investigating the local environment, the children can consider the bearings (from the school) of a range of local amenities.

Using the Logo computer environment is a valuable way of giving a dynamic element to work in this area. It allows a basic understanding of position and movement to be reinforced, and provides opportunities to gain new insights.

BY THE END OF Y5/P6, MOST CHILDREN SHOULD BE ABLE TO:

- read and plot coordinates in the first quadrant, using the terms 'x-axis' and 'y-axis'
- recognize perpendicular and parallel lines.

BY THE END OF Y6/P7, MOST CHILDREN SHOULD BE ABLE TO:

- recognize positions and directions, and start to use bearings (particularly in geography work)
- read and plot coordinates in all four quadrants
- recognize perpendicular and parallel lines in quadrilaterals
- use the terms 'intersecting lines' and 'intersection'.

SOME COMMON MISCONCEPTIONS AND STRATEGIES FOR CORRECTING THEM

PARALLEL LINES

Children often fail to identify parallel lines that are in oblique orientations or varying distances apart:

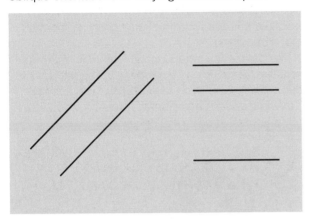

Present a range of sets of parallel and non-parallel lines, and challenge the children to explain which are parallel and why.

NEGATIVE COORDINATES

Children often have difficulty with reading coordinates in four quadrants, where the numbers to the left of 0 horizontally and below 0 vertically are negative. The quadrants themselves are labelled around in an anti-clockwise direction (see diagram below). When plotting a point such as (5, –3) or (–2, 6), some children will ignore the '–' and plot the points in the first quadrant.

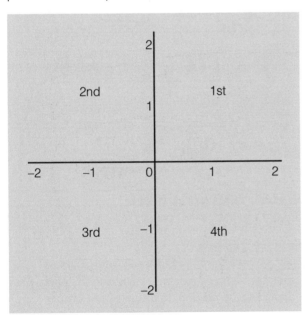

Revise work on negative numbers, using both horizontal and vertical number lines, before working with four quadrants. The coordinate grid can be explained as the juxtaposition of a horizontal and a vertical number line.

THE WAY FORWARD

When Logo is used, the orientation of the computer screen means that 'forward' is interpreted as 'up' and 'back' as 'down'. This can be confusing for children who are switching between the computer screen and a floor robot.

When entering commands in Logo, the children should try to imagine that they are the turtle, and consider the movements relative to their own position. Work with the Valiant Roamer is useful for developing this ability.

Further confusion may arise when children go from a computer environment to map work involving real locations. Many children fail to realize that the direction 'N' on a map does not necessarily correspond to 'forwards'. Having a compass to refer to is useful in such work.

POSITIVE AND NEGATIVE

†† *Whole class, then individuals*
🕑 *50 minutes*

AIM
To use coordinates in more than one quadrant.

WHAT YOU WILL NEED
Squared paper, photocopiable page 50, an OHT or enlarged version of the grid on page 50, a set of 21 cards marked from –10 to 10 (including 0), OHP.

WHAT TO DO
Start by revising the use of negative numbers – for example, by counting around the class backwards from 30 in twos or threes using a –50 to +50 number line. Now introduce the four-quadrant diagram. One way of thinking of this diagram is to consider it as formed by a pair of perpendicular number lines, crossing at 0 on each.

Show the children the enlarged or OHT version of the diagram on page 50. Explain the labelling of the four quadrants (see 'Key ideas', page 42). Check that the children know that the first coordinate of a pair represents the position on the horizontal axis (or x-axis), and the second represents the position on the vertical axis (or y-axis).

Shuffle the pack of number cards and invite a child to pick two cards to represent the x and y positions respectively. Invite another child to plot this point on the diagram. Ask the rest of the class to check. Repeat this several times. If no problems arise, volunteer to try some yourself and make some deliberate errors – can the children correct you?

Now draw a rectangle or triangle clearly on the grid. Make sure that it straddles either the x or the y axis, so that the coordinates will include a mixture of positive and negative numbers. Ask the children to discuss with their neighbours what the coordinates of the corners of the shape are; then check through with the whole class. Repeat this two or three more times. Now distribute copies of page 50 and give the children time to work through it individually.

In a plenary session, work through the answers to page 50: a) square, b) oblong, c) right-angled triangle, d) pentagon, e) parallelogram, f) obtuse-angled triangle. Ask for volunteers to read out the coordinates of their own shapes, so that these can be checked through by everyone else.

DISCUSSION QUESTIONS
● *Who can remember the rules for plotting coordinates?*
● *How can you tell which quadrant any coordinate pair will lie in?*

ASSESSMENT
Can the children successfully place coordinates in all four quadrants?

EXTENSIONS
● The children can use another copy of the grid from page 50 (or a sheet of suitably labelled squared paper) to create a more extensive coordinate design (such as a house, boat, rocket or face), using each of the quadrants.
● Photocopiable page 51 provides further individual practice (perhaps as homework), linking the use of coordinates to work on reflecting shapes (see the 'Transformations and symmetry' section, page 28).

DOUBLE TIME

†† *Whole class, then individuals*
🕑 *50 minutes*

AIM
To investigate the effect of doubling coordinates.

WHAT YOU WILL NEED
Squared paper, rulers, pencils.

WHAT TO DO
First, play a warm-up coordinate game with the whole class (see ideas in 'Plot the shape', page 49).

Now ask the children to prepare their own coordinate grid, with 0–10 on each axis, using squared paper. On the board or flip chart, write up the following pairs of coordinates: (1, 1), (1, 3), (3, 5), (5, 3) and (5, 1). Ask the children to plot these and join them up in order. This should produce a drawing like the one shown opposite.

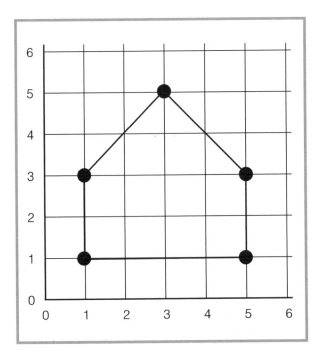

ASSESSMENT

Can the children predict the effects on a shape of doubling either (or both) of the x and y coordinate values? Can they calculate the new coordinates and plot them successfully?

VARIATION

● The children can start with a large coordinate drawing and experiment with halving either the x or the y coordinates.

EXTENSIONS

● The children can use the writing frame on resource page 128 to reflect on the results of this activity.
● Children who have already worked in four quadrants (using negative as well as positive coordinate values) can try this activity using the grid on photocopiable page 50. This will extend their experience of negative coordinates, and will also provide a context for multiplying negative numbers.
● Photocopiable page 52 gives further individual practice in doubling coordinates and seeing the effects. It could be used for follow-up homework or further class discussion. The children can use squared paper to draw the shapes, or use their visualization skills to sketch them.
● The activity 'Double up' (page 84) considers in a more formal way the effects on a shape's area of doubling the length of its sides.

Now ask the children what the coordinates would be if each of the x-axis values **only** were doubled. They should suggest the following: (2, 1), (2, 3), (6, 5), (10, 3) and (10, 1). Ask them to imagine what shape this new set of coordinates will produce. After discussing their ideas, let them plot the points (on the same grid, perhaps using a different colour) to discover what happens. The result should be a wider version of the original shape.

Now ask the children what the coordinates would be if each of **only** the y-axis values of the original shape were doubled. They should suggest the following: (1, 2), (1, 6), (3, 10), (5, 6) and (5, 2). Again, ask them to imagine what shape this new set of coordinates will produce. After discussing their ideas, let them plot the points to discover what happens. The result should be a taller version of the original shape.

Give out more squared paper and ask the children to draw another pair of 1–10 axes. Now ask them to create their own coordinate shape or pattern. They should make a note of the coordinates they have used, then experiment with doubling the x or y value, doubling both, or even trebling either. Some of these experiments should produce interesting and attractive visual effects, particularly when a different colour is used for each version of the shape; these can be shown and discussed in a plenary session. The children can also display their drawings and challenge the class to work out what they have done to the coordinates.

DISCUSSION QUESTIONS

● *Can you imagine what that shape will look like if you double the x [or y] coordinates?*
● *What has happened to the coordinates for that shape?*

POSITION & DIRECTION

SLUG TRAILS

👥 *Whole class, then individuals*
🕐 *50 minutes*

AIM
To look at patterns based on a repeated turn and movement.

WHAT YOU WILL NEED
Squared paper, rulers, pencils, protractors, the story of Sly Slug (see below) as an OHT, or written up on the board or flip chart.

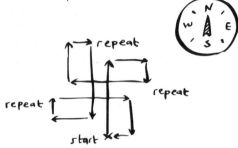

Sly Slug starts off on his travels facing north. On the first day he travels 4cm, then turns 90° right before resting. On the second day he travels 2cm, then turns 90° right before resting. On the third day he travels 1cm, then turns 90° right before resting. He then repeats this whole sequence of movements several times.

WHAT TO DO
Start by rehearsing the vocabulary the children will be using: forward, back, right, left, 90° and the compass directions. Now display the OHT with the story of Sly Slug.

Ask the children to recreate the journey of Sly Slug. Distribute squared paper, and tell them to start at the centre of the paper. They should draw a compass in the corner of the sheet, so that they can keep track of which direction he is facing at any time. For this initial pattern, they should find that after completing the sequence of turns and movements four times, Sly Slug is back where he started, facing north (see illustration below). You may want to lead the class through this example, checking that everyone has followed the sequence carefully.

Now suggest that the children investigate what happens when the length of the journey changes – for example, 3cm on the first day, 1cm on the second and 2cm on the third, still turning 90° right at the end of each day and repeating the sequence every three days. The illustration below shows the pattern created. Depending on the choice of lengths, a range of repeating patterns can be made. Discuss the different results obtained.

DISCUSSION QUESTIONS
● *Which direction is Sly Slug now facing?*
● *What do you think will happen if you change some of the distances that Sly Slug travels?*
● *Can you explain how that pattern came about?*
[The slug has moved the same distance in each of four directions: the movements north/south and east/west cancel each other out.]

ASSESSMENT
Can the children recreate Sly Slug's journey? Can they change the conditions of the journey and comment on the results?

EXTENSIONS
● The children can recreate Sly Slug's journey using Logo. Depending on the version used, it might be sensible to multiply each length by 10. The value of using Logo for this investigation is that it can deal with the repeat instructions very efficiently. Try using: Repeat 4[fd 40 rt 90 fd 20 rt 90 fd 10 rt 90] (The distances have all been multiplied by 10 because of the small size of standard Logo movements.) Once the procedure has been created and saved, it can easily be edited to try out new ideas.
● The children can explore what happens when they vary the angle of turn, perhaps trying with 45° or 30°. This is best done using plain paper and a protractor.
● This work could be further extended to include simple examples of **bearings** – for example, using

'030°' to mean a turn of 30° clockwise and '330°' to mean a turn of 30° anticlockwise.
● They can use the writing frame on resource page 128 to reflect on the results of this investigation.

PARALLEL OR PERPENDICULAR?

†† *Whole class, then pairs*
⏲ *50 minutes*

AIM
To investigate parallel and perpendicular lines.

WHAT YOU WILL NEED
Resource page 121 (one copy per pair), photocopiable page 53 (one copy per child).

WHAT TO DO
Write the words 'parallel' and 'perpendicular' on the board or flip chart. Ask the children whether they know what these words mean. Suitable definitions to draw out are as follows.
● **Parallel lines** are ones that never meet. Take a ruler and, without moving it, draw a line on either side of it. Ask the children to imagine what would happen if the two lines were continued. Good examples of parallel lines are railway lines, lines in an exercise book and parallel bars in a gym. Show the children the convention of marking pairs of parallel lines with small arrows:

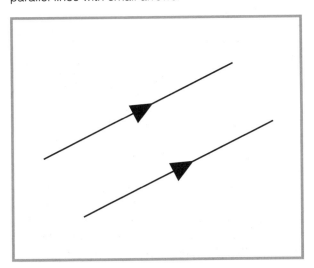

Discuss the idea that parallel lines do not have to be horizontal or vertical, and that several parallel lines do not have to be the same distance apart (see page 43 for examples).
● **Perpendicular lines** are lines that meet at right angles. Good examples of perpendicular lines are squared paper, window frames and other square-based grilles. In these cases, the perpendicular lines cross completely; in other cases, they do not, but the perpendicular relationship still exists – for example, the corner of a square or of a table.

When the children are happy with the idea of parallel and perpendicular lines, distribute one copy per pair of resource page 121. Ask them to find (and agree on) all the parallel and perpendicular lines on each quadrilateral. They should mark perpendicular lines by putting a large 'P' in the appropriate corner, and mark parallel lines using arrows. If the shape has more than one pair of parallel lines (as in a parallelogram), they should mark the second pair with two little arrows. Bring the class back together to review their findings.

Now introduce photocopiable page 53, which introduces a wider range of shapes. The children have to mark the parallel and perpendicular lines as before, then draw shapes to meet given criteria. Depending on the time available, they can complete this in the lesson or as homework.

DISCUSSION QUESTIONS
● *Where can you find examples of parallel lines?*
● *Where can you find examples of perpendicular lines?*
● *Why can't a triangle have parallel sides?*

ASSESSMENT
Can the children identify perpendicular and parallel lines in real-life objects around them (Y5/P6), and in a range of shapes (Y6/P7)? Can they describe the difference between parallel and perpendicular lines?

EXTENSION
● The children can make a poster to show parallel and perpendicular lines in real-life objects.

POSITION & DIRECTION

USING LOGO

†† Whole class, then pairs
🕐 50 minutes

AIM
To make and recreate patterns using Logo.

WHAT YOU WILL NEED
A Logo computer program, resource page 127.

WHAT TO DO
Your starting point will depend on the children's previous experience with Logo. This activity assumes that they already have some basic knowledge of the program and are aware of the functions of the forward, backward, right, left and repeat commands. Even if the children have used Logo before, it is worth demonstrating these commands – check in your Logo manual for the exact commands for your version, and make a poster of these for display.

Give out one copy per child of resource page 127, and choose a pattern to make together. A good way to encourage children with Logo is to get them to trace their finger over the pattern, thinking about the directions they will need to turn at different times. Ask the children to suggest commands that you can enter into the program to recreate the chosen pattern. At first, this might develop through trial and improvement; but encourage the children to become more efficient in their program writing, especially through the use of the repeat command.

Depending on the availability of computer time, organize a rota to let the children (working in pairs) try to recreate one of the designs on screen and print it out. Different pairs can be allocated different shapes, so that all the shapes can be displayed. They can use the computer to design similar patterns for their friends to recreate.

DISCUSSION QUESTIONS
● *What would be the best way to start the pattern?*
● *What commands do we need to use?*
● *Is that the best way to make that pattern?*

ASSESSMENT
Can children use Logo to recreate a simple pattern? Can children describe the commands they used, explaining what each one does? Can children use the repeat command in order to program more efficiently?

EXTENSIONS
● The children can print off the commands they used to recreate the pattern, and write an explanation of what each command does. This work can be displayed alongside their patterns.
● Use a range of multicultural patterns, from Roman mosaics to Islamic designs, to inspire the children's programming.

PLOT THE SHAPE

👥 *Whole class, then pairs*
🕐 *50 minutes*

AIM
To revise the use of coordinates in the first quadrant.

WHAT YOU WILL NEED
Resource page 126, pencils, dice (two per pair in different colours).

WHAT TO DO
Draw a 6 × 6 coordinate grid (see figure below) on the board or flip chart, or use page 126 as an OHT. Remind the children of the rules for plotting coordinates: the horizontal point is always given first in a coordinate pair, so the point (3, 4) refers to three across and four up.

Ask children to come to the front and plot pairs of coordinates. Ask the class to say whether each point is correct. Alternatively, you could play the devil's advocate: ask a child to give you a pair of coordinates, then plot them wrongly. For example, if the child says (2, 5), plot (5, 2) or mark the point in the middle of a square rather than at the intersection of two lines. Can the children tell you what's wrong?

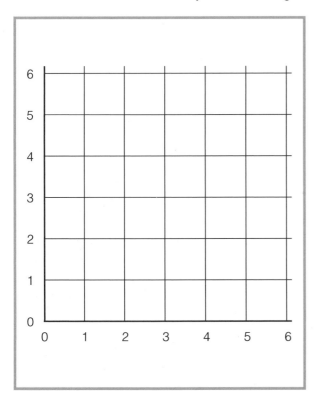

Now introduce the 'Plot the shape' game – the best way to do this is as a whole-class demonstration, with one child playing against you. You need two different-coloured dice: one represents the horizontal coordinate, the other the vertical coordinate (it might be worth marking which is which on the board, flip chart or OHT). The rules are as follows:

> 1. The two players take turns to throw the dice and plot the points. The aim of the game is to create a particular shape by joining up some of the points plotted. The shape should be agreed beforehand – try a square, oblong or right-angled triangle to begin with.
> 2. If a pair of points has already been plotted, the player throws again (as many times as necessary) until a new pair comes up.
> 3. After each turn, the players should look at the grid and see whether it is possible to join points to make the required shape.
> 4. The player whose coordinate first completes the shape chooses the shape for the next game.

The children should pair up and play on a copy of page 126, which provides grids for six games.

In a plenary session, reflect on the rules of the game: could they be changed to make it harder or easier? For example:
● If the point has already come up, the player forfeits his or her turn.
● Use a larger grid and a pair of 0–9 spinners.
● Start with several shapes already drawn on the grid. The players claim the corners of the shapes as they throw the coordinates. The player who claims the final corner wins the shape.

DISCUSSION QUESTIONS
● *How do you know where to plot (4, 2)?*
● *What's the difference between plotting (2, 5) and (5, 2)?*
● *What did you like about the game?*
● *Which shapes were easiest to make? Which were more difficult?*

ASSESSMENT
Do the children recall the rules for plotting coordinates? Can they identify points on the grid that can be joined to create a particular shape?

EXTENSIONS
● The players can use two identical dice and choose which number to use for each coordinate (horizontal or vertical).
● They can use squared paper to create a 10 × 10 coordinate diagram (numbered 0–9), and use sets of digit cards to generate random coordinates.
● Photocopiable pages 54 and 55 provide further individual practice in identifying the coordinates of points in the first quadrant. The answers for page 54 are: squares (2, 2) and (3, 5); rectangles (4, 1) and (5, 2); parallelograms (6, 1) and (4, 6).

NAME   DATE

THE BIG GRID

■ Plot these sets of points on the big grid.

■ What shape have you made with each set of points?

A. (–7, 2), (–7, 4), (–5, 2), (–5, 4)   B. (1, –1), (1, –5), (3, –1), (3, –5)

C. (–8, –9), (–6, –9), (–8, –5)   D. (–3, 1), (–3, 4), (0, 6), (3, 4), (3, 1)

E. (–2, –8), (–1, –6), (5, –6), (4, –8)   F. (–4, 9), (4, 7), (0, 7)

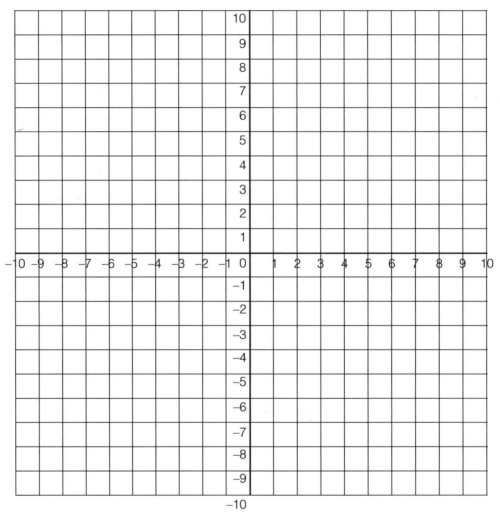

■ Draw four more shapes, one in each quadrant. Write the coordinates here:

G. _____   H. _____

J. _____   K. _____

 How can you tell, by looking at the coordinates, which quadrant each point is in?

SEE 'POSITIVE AND NEGATIVE', PAGE 44.

NAME

DATE

REFLECT THE SHAPE

■ Reflect shapes A to D in the y-axis and E to G in the x-axis.

A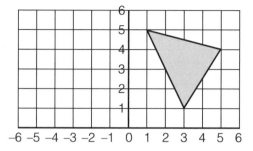

B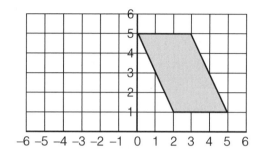

C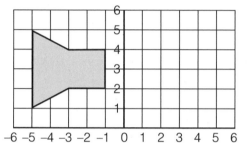

D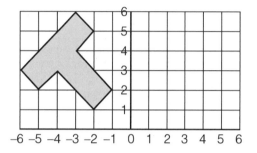

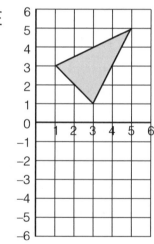 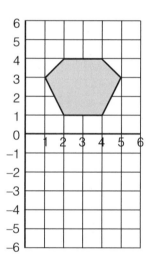 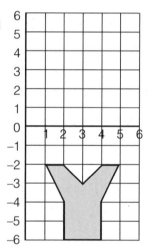

■ Record the coordinates of the corners of the reflected shapes on a blank sheet of paper.

 Check your answers with a friend. Why were some reflections easier to work out than others?

SEE 'POSITIVE AND NEGATIVE', PAGE 44.

DEVELOPING SHAPE,
SPACE & MEASURES

POSITION & DIRECTION

DOUBLE UP

1. Imagine a drawing made by joining these three coordinates:
   (1, 1), (1, 5), (3, 1).
   Now double the **x** values.
   The new coordinates will be:
   Now sketch the original shape and the new shape. You can use a sheet of squared paper to help you, or just rely on what you can visualize.

   Original   **New**
   shape   **shape**

2. Imagine a drawing made by joining these four coordinates:
   (2, 1), (3, 2), (6, 2), (5, 1).
   Now double the **y** values.
   The new coordinates will be:
   Now sketch the original shape and the new shape.

   Original   **New**
   shape   **shape**

3. Imagine a drawing made by joining these four coordinates:
   (1, 0), (0, 2), (2, 3), (3, 1).
   Now double both the **x** and the **y** values.
   The new coordinates will be:
   Now sketch the original shape and the new shape.

   Original   **New**
   shape   **shape**

 Compare your answers to a friend's. Do you agree on which shape was easiest to visualize, and which was hardest?

SEE 'DOUBLE TIME', PAGE 44.

PERPENDICULAR OR PARALLEL?

■ Mark all the perpendicular lines and parallel lines on each of these shapes.

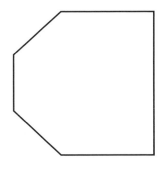 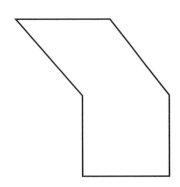 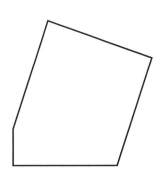

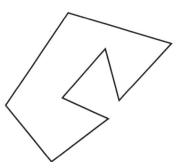 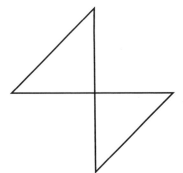 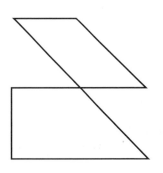

■ Now try to draw the following:

1. A shape with two pairs of parallel lines, but only one pair of perpendicular lines.	2. A shape with at least two pairs of each type of line.	3. A shape with at least two pairs of perpendicular lines, but no parallel lines.

 On the back of this sheet, in your own words, explain the difference between parallel and perpendicular lines.

SEE 'PARALLEL OR PERPENDICULAR?', PAGE 47.

DEVELOPING SHAPE,
SPACE & MEASURES

POSITION & DIRECTION

NAME

DATE

MISSING CORNERS

■ Plot a fourth point to complete each shape. Write the point's coordinates below.

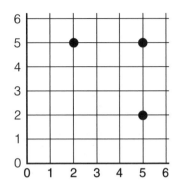

A square _____

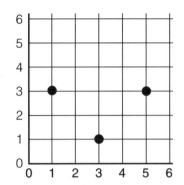

A square _____

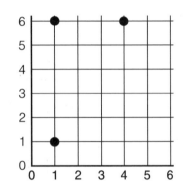

A rectangle _____

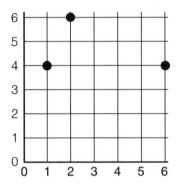

A rectangle _____

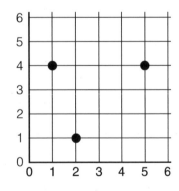

A parallelogram _____

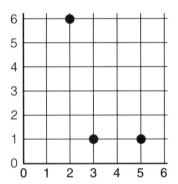

A parallelogram _____

 Use a blank coordinate grid sheet to make up some more puzzles like these. Swap with a friend: can you solve each other's puzzles?

SEE 'PLOT THE SHAPE', PAGE 49.

POSITION & DIRECTION

NAME _____   DATE _____

COORDINATE DESIGNS

■ In each case, draw the design required, then write down the coordinates of the points you have used.

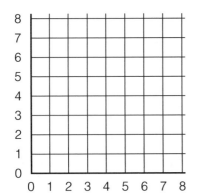

A large octagon

Coordinates used: _____

_____

_____

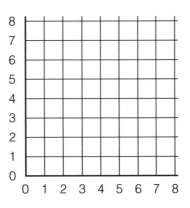

Your initials

Coordinates used: _____

_____

_____

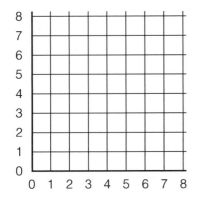

A house

Coordinates used: _____

_____

_____

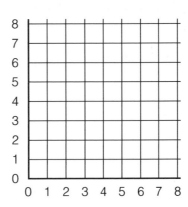

Your own design

Coordinates used: _____

_____

_____

 Check the coordinates for your own design, then write them out on plain paper.

 Can your friend use them to make the same design on squared paper (or a blank copy of this sheet)?

SEE 'PLOT THE SHAPE', PAGE 49.

DEVELOPING SHAPE,
SPACE & MEASURES

POSITION & DIRECTION

KEY IDEAS

- Classifying angles formally.
- Estimating angles and measuring them accurately.
- Knowing the angle properties of shapes and straight lines.
- Using a protractor.

The idea that angle is a measure of turn needs to be constantly reinforced. Children tend to see angle in a very static way, perhaps due to an overemphasis on the corners of shapes. The idea of angles as amounts of turn can be reinforced by referring to the hands on a clock face or a person turning to face different compass directions. (Note that bearings are covered as an extension of work on 'Position and direction' – see page 42.) Being able to estimate angles is also important; this aspect parallels the teaching of other measures, such as length and mass.

Before the children start using a protractor to measure angles, they need to be able to describe the differences between acute, right, obtuse and reflex angles with confidence, since these are used as 'marking posts' to check that an angle's size has been measured correctly. When the children are using a protractor for the first time, it is best to measure to the nearest 5° only; measuring to the nearest degree using a protractor (like using a ruler

to measure to the nearest millimetre) can be quite difficult.

Another aspect of angle work that is new at this stage is the use of formally described angle properties to calculate missing angles. For example, if two angles of a triangle are known, the third can be found by subtraction from 180°. To ensure that this is not just a meaningless calculation exercise, the children need to be given opportunities to discover such angle properties in a practical context first.

BY THE END OF Y5/P6, MOST CHILDREN SHOULD BE ABLE TO:

- identify, estimate and order acute and obtuse angles
- understand and use angle measure in degrees
- use a protractor to measure and draw angles to the nearest 5°
- calculate angles on a straight line.

BY THE END OF Y6/P7, MOST CHILDREN SHOULD BE ABLE TO:

- recognize and estimate angles, including reflex angles
- use a protractor to measure and draw acute and obtuse angles to the nearest degree
- calculate angles in a triangle or around a point
- explore angles using appropriate ICT.

Encourage the children to estimate first, and to check by referring to whether the angle to be measured is acute or obtuse.

Finally, it is important to keep the protractor in line with the base of the angle. If this problem occurs, the child is not seeing the 'turn' within the corner:

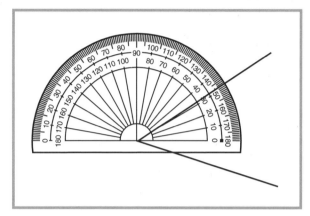

Ask the child to place a finger along the end of one arm of the angle, then turn it to match the other arm. Stress that this is what is being measured. The child can also draw an arrow inside the angle to represent the turn.

SOME COMMON MISCONCEPTIONS AND STRATEGIES FOR CORRECTING THEM

PROTRACTED PROBLEMS

Several difficulties can arise from using a protractor. They reveal misconceptions about the nature of angle, as well as the mechanical difficulty of using a protractor. Recognizing this is a key to encouraging the confident measuring of angles.

The first potential problem is where to place the protractor. Here, the child has confused using a protractor with using a ruler:

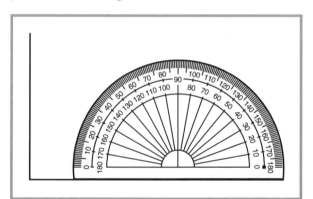

Remind the child what is being measured: the angle, not the length.

The next potential problem is deciding which way round to use the scale. For example, is the angle shown below 50° or 130°?

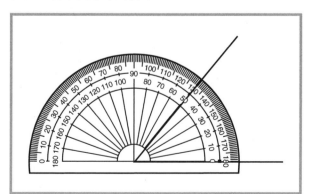

DEVELOPING SHAPE, SPACE & MEASURES

ANGLES

AROUND AND AROUND

†† Whole class
🕐 15 minutes, repeated several times

AIMS
To reinforce the idea of angle as an amount of turn. To appreciate the relative size of a whole turn, a half-turn and a quarter-turn. To relate parts of a turn to numbers of degrees.

WHAT YOU WILL NEED
Space to move around, a poster or OHT showing the diagrams below:

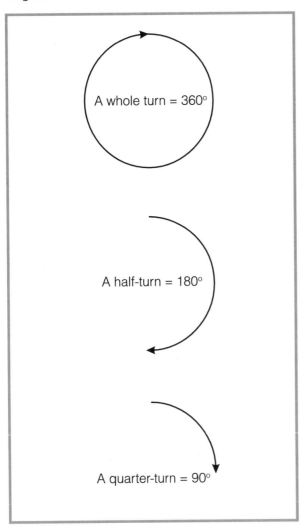

A whole turn = 360°

A half-turn = 180°

A quarter-turn = 90°

WHAT TO DO
This activity is best carried out in the school hall. Display the poster or OHT for the children to refer to (as they become more confident, it can be taken out of sight). Ask the children to stand up facing you. Remind them what it means to make a whole turn (you turn on the spot until you are back where you started), a half-turn either left or right (you end up

facing the opposite direction) and a quarter-turn (here the direction of turn affects the outcome). Let the children try out these different amounts of turn. Discuss the effects of turning in different directions.

Now explain that there are 360 degrees in a full turn, and so we can say '360 degrees' instead of saying 'whole turn'. Ask the children:
● *How many degrees are there in a half-turn?*
● *How many degrees are there in a quarter-turn?*
Now tell the children that you are going to call out different angles, and they should turn to the right the amount that you say. Explain that you do not expect them to be able to make the **exact** turn – but they should give a fair impression of how far the turn gets them. For example, if you say 70°, they should turn nearly a quarter-turn; 110° should be slightly more than a quarter-turn; 140° should be about half-way between a quarter-turn and a half-turn.

After each turn, check that everyone has understood. If there is confusion, choose someone to demonstrate the correct turn. After each turn, the children should return to face the original direction before you call out another turn. In this session, stick to turns in one direction.

In a subsequent session, try with left turns. In later warm-up activities (to start PE lessons), mix up left and right turns. You could try this as a 'Simon says' game.

DISCUSSION QUESTIONS
● *Where will you be facing after a quarter-turn/a half-turn/a whole turn?*
● *Why does [for example] a 140° turn leave you facing that way?*

ASSESSMENT
Can the children (judging by their responses to instructions) recognize a full turn as being equal to 360°, a half-turn as being 180° and a quarter-turn as being 90°? Can they make sensible estimates of the different turns as they follow the instructions?

EXTENSIONS
● Introduce compass directions into the game. If you start by facing north, a turn of 130° to the right leaves you facing somewhere between east and south, while a turn of 250° to the right leaves you facing somewhere between south and west. Can the children identify the approximate compass directions they are facing, using first four and then eight compass points? (For more work on bearings, see the section on 'Position and direction', page 42.)
● Introduce the terms 'acute' (angles less than 90°), 'obtuse' (angles more than 90° but less than 180°) and 'reflex' (angles more than 180° but less than 360°). Use these terms to characterize the turns you ask the children to make. For example, you might say: *Turn 70°, an acute angle.* Ask the children to confirm the name of each angle they turn.

DIFFERENT ANGLES

†† *Whole class, then individuals*
🕐 *50 minutes*

AIM
To revise the names of different angles (acute, right, obtuse, reflex, straight line).

WHAT YOU WILL NEED
Resource pages 120 and 121 (enlarged or copied onto OHTs), the 'Shape and space glossary' on pages 8–9 (enlarged or copied onto an OHT), photocopiable page 66.

WHAT TO DO
Discuss the terms used to describe different types of angle: 'acute', 'right', 'obtuse', 'reflex', 'straight line'. (You could display an enlarged or OHT version of the 'Angles' section in the 'Shape and space glossary'.) A key distinction, particularly when looking at shapes, is that between acute, right and obtuse angles. If the children are unsure of this, you could revise the method of making a right angle from a scrap of paper: folding it in half and then in half again produces a right angle. Angles smaller than this are acute; angles greater than this (but not greater than a half-turn) are obtuse.

Ask the children to look at the triangles and quadrilaterals on the enlarged or OHT versions of pages 120 and 121. Ask individuals to tell you whether particular angles are acute, right, obtuse or reflex. (The chevron has a reflex angle.)

Once the children are comfortable with the vocabulary, distribute copies (one per child) of page 66: the 'Angles quiz'. In Section A, the children should categorize angles given as degree measures. In Section B, they should order the sizes of angles described in a variety of ways. Some children will benefit from being encouraged to draw rough sketches of the different angles in Section B; more confident children should try to visualize them.

In a plenary session, review both sections of the 'Angles Quiz'. The answers are: Section A 1. obtuse, 2. right, 3. reflex, 4. acute, 5. reflex, 6. straight line, 7. obtuse, 8. reflex, 9. obtuse. Section B 1. 89°, a right angle, a half-turn, 200°. 2. 75°, a quarter-turn, 180°, a reflex angle. 3. an acute angle, 90°, 175°, a straight line.

DISCUSSION QUESTIONS
● *How can you tell whether or not an angle is acute/obtuse/reflex?*
● *Which types of angle do you find easiest to recognize? Which are more difficult?*

ASSESSMENT
Can the children use the different angle terms with confidence, assigning each term to the appropriate range of angle sizes? Can they order a set of angles described using a range of terms?

EXTENSION
The children can make up their own 'Angle Quiz' sheet, using page 66 as a template, then swap sheets with a friend and work out the answers.

50

DEVELOPING SHAPE,
SPACE & MEASURES

ANGLES

MEASURING ANGLES

†† *Whole class*
🕐 *50 minutes*

AIM
To estimate and measure angles.

WHAT YOU WILL NEED
Photocopiable page 67 and/or page 68, protractors, pencils, geostrips (two per child).

WHAT TO DO
Start the session with a quick oral warm-up on the names of angles, using questions similar to those on page 66.

Spend some time demonstrating how to use a protractor. Draw an angle on an OHT, or on the board or flip chart, and measure it. An OHP is ideal, since an ordinary transparent protractor will show up on it quite clearly. There are potential difficulties in using the standard 180° semi-circular protractor, and your demonstration should highlight these (see 'Key ideas', pages 56–57). The previous two activities in this chapter would help the children to have a sense of the relative numbers of degrees in different angles , and an understanding that angle is a measure of amount of turn.

These ideas should be reiterated in your demonstration. The key points in instructing children to use a protractor for measuring angles are:

1. The base line of the protractor should be placed over one of the 'arms' of the angle.
2. The centre point of the protractor should be placed at the point where the two arms of the angle meet (see Figure A).
3. What is being measured is the amount that one arm needs to be turned so that it is over the other arm. (This can be demonstrated using a pair of geostrips: make an angle with them; then, keeping one still, turn the other until it is over the first. This amount of turn represents the angle.) Imagining one arm turning until it is over the other should help the children to decide which way round the protractor's scale should be used. They can draw a little arrowed arc between the arms of an angle, showing which direction they are measuring it in (see Figure B).
4. Encouraging the children to state whether an angle is acute or obtuse should also help: if an angle can be read as either 50° or 130° (because the child is unsure which scale to use), then recognizing that the angle is either acute or obtuse should help the child to make the right choice (see Figure C).

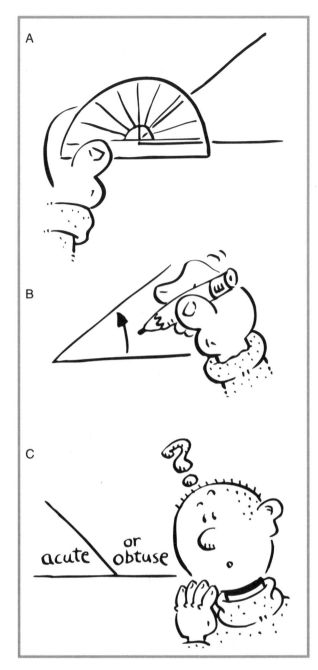

A

B

C

acute *or* obtuse

It is also important to encourage the children to estimate, just as you have done when introducing the measurement of length or mass.

The children's previous experience should help you to decide how basic the instruction should be. However, do not take too much for granted: even if the children have used a protractor before, they are unlikely to have used it outside a maths lesson, and so may be out of practice.

After discussing the use of a protractor, let the children work individually to estimate and measure the angles on photocopiable page 67 and/or page 68. On page 67, all the angles have one horizontal 'arm' and the answers are all multiples of 5°. Page 68 has angles in a variety of orientations; these should be measured to the nearest 1° (or 5° if that is more appropriate). The emphasis should be on slow, careful work; children who have finished and checked their answers could help others.

The answers to page 67 are 1. 40°, 2. 75°, 3. 130°, 4. 25°, 5. 30°, 6. 120°, 7. 155°, 8. 50°. The answers to page 68 are: 1. 40°, 2. 109°, 3. 34°, 4. 141°, 5. 77°, 6. 114°, 7. 47°, 8. 19°. (Answers to within 1° should be considered correct here. Allow answers to the nearest 5° for younger or less confident children.)

DISCUSSION QUESTIONS
● *How can you tell that this angle is less than/more than 90°?*
● *What are good ways to be sure that you are using your protractor correctly?*

ASSESSMENT
How well do the children understand how to use a protractor (ie can they explain the potential difficulties)? How accurately do they measure the angles?

VARIATION
The children can create their own pages of angles for their friends to estimate and measure.

EXTENSION
Introduce the children to a 360° angle measurer.

DRAWING ANGLES

†† *Whole class, then individuals*
🕐 *50 minutes*

AIM
To draw angles of different sizes accurately.

WHAT YOU WILL NEED
Protractors, sharp pencils, rulers, blank paper, photocopiable page 69.

WHAT TO DO
Start with a quick oral warm-up, as in 'Measuring angles' (page 60): check that the children can name different-sized angles and know the relative sizes of the different kinds of angle.

Recap the key things to remember when measuring angles with a protractor. Now demonstrate to the class how to use a protractor to draw an angle of any given size:
1. Start with a straight horizontal line.
2. Place the protractor at one end, with its base line along the line and its centre at the end of the line.
3. Mark a point on the paper at the edge of the protractor, corresponding to the angle required. Use knowledge of the difference between acute and obtuse angles to check that you are using the right scale.
4. Use a ruler to join a line from the end of the original line to the point marked.

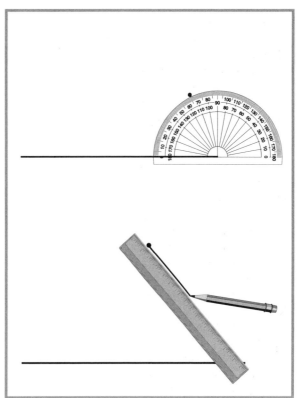

DEVELOPING SHAPE, SPACE & MEASURES

Now let the children have a go. Ask them to draw five or six angles each. Choose angles that are simple multiples of 5° to begin with. They can check each other's work. After a while, stop the class and reiterate the angle drawing procedure. Children struggling with this should repeat it until they are confident, or stick to drawing angles that are multiples of 10°. Those without problems should go on to try each of the extensions listed below.

DISCUSSION QUESTIONS
● *Who can tell me how to use the protractor for drawing angles?*
● *What is the difference between an acute angle and an obtuse angle?*
● *What helps you to draw angles accurately?*
● *How can you check that you have drawn the angle correctly?*

ASSESSMENT
How accurately do the children draw angles? How well can they describe the process of drawing an angle?

EXTENSIONS
● Ask the children to start with an oblique line rather than a horizontal line.
● Suggest angles to draw that are not simple multiples of 5°.
● The children can draw an angle at each end of the line, then continue the two oblique lines until they meet (at least one will have to be an acute angle for this to work), then measure the angle between them. They may notice that the three angles add up to a straight line (see 'Angles in a shape', page 63).
● Photocopiable page 69 can be used for further consolidation of this topic.

ANGLES IN A SHAPE

👥 *Whole class, then groups*
🕐 *50 minutes*

AIM
To look at the result of adding up the angles in a triangle and in a quadrilateral.

WHAT YOU WILL NEED
Resource page 120 and/or page 121, blank paper, pencils, rulers, a large paper triangle and/or quadrilateral, Blu-tack or an OHP.

WHAT TO DO
You may prefer to split up the content of this session, looking at triangles and quadrilaterals in separate sessions. The instructions that follow use a triangle as an example; where the result for a quadrilateral is different, this is also noted.

Hold up a large paper triangle and tell the children that you want to add up all the angles. Invite three children to tear off a corner each, then show the class how these three corners can be arranged to make a straight line (see illustration below). This can be demonstrated by reassembling the three torn corners on an OHP, or by sticking them onto the board with Blu-tack.

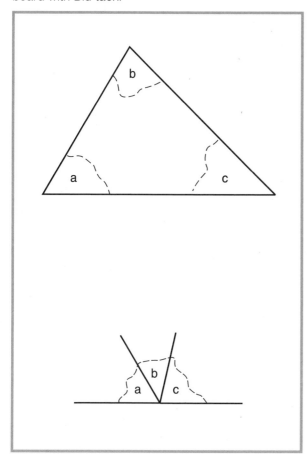

The three angles of the triangle form a straight line, and therefore add up to 180°. If you are doing this with a quadrilateral, the four angles will form a whole turn or 360°:

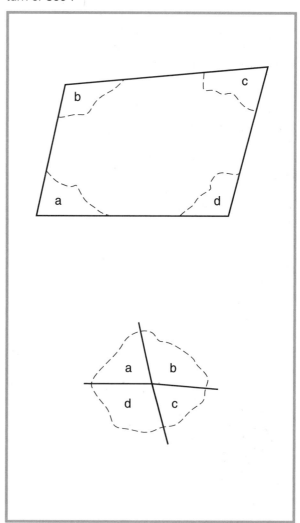

But what if you have just chosen a 'special triangle' that this works for? Invite the children to try it for themselves with different triangles. They should each draw a large triangle (using up most of an A4 sheet), tear off the corners and stick them together on another sheet of paper. Encourage some children at each table to try with right-angled and obtuse-angled triangles.

A second way of showing that the angles of a triangle add up to 180° is to measure the angles in lots of triangles and add them up. Resource page 120 can be used for this. Remind the children to measure carefully. Sometimes a total of 179° or 181° will occur – this is because the angles are measured to within 1° of accuracy, which can lead to slight errors. Resource page 121 can be used in a similar way to show that the angles of a quadrilateral always add up to 360°.

When the children have measured the angles in all of the triangles and/or quadrilaterals on these sheets, they can draw some more of their own to check.

63

ANGLES

DISCUSSION QUESTIONS
● *Do you think this will work for all triangles/ quadrilaterals?*
● *Did the results surprise you?*
● *If you know two angles of a triangle, how can you find the third one without measuring?*
● *If you know three angles of a quadrilateral, how can you find the fourth one without measuring?*

ASSESSMENT
Can the children explain in their own words what they have learned about the angles of a triangle/ quadrilateral? Can they measure the angles in a triangle/quadrilateral accurately, and add them up correctly?

EXTENSION
The children can use the writing frame on resource page 128 to reflect on their findings.

ANGLE CHALLENGE

†† *Whole class, then individuals*
🕐 *40 minutes*

AIM
To calculate missing angles, using knowledge of angle properties.

WHAT YOU WILL NEED
Photocopiable pages 70 and 71, pencils, calculators (optional).

WHAT TO DO
The two photocopiable sheets can be used as the basis of separate sessions, or together in one longer session. To introduce page 70, remind the children that a half-turn is equivalent to 180° and a whole turn to 360°. Draw some examples similar to those on page 70 on the board or flip chart, and ask the children how they could calculate the missing angles.

Page 71 is concerned with the angle sums of triangles and quadrilaterals that the children have considered in the activity 'Angles in a Shape' (page 63). Remind them that the angles in a triangle always add up to 180°, while the angles in a quadrilateral always add up to 360°. Again, draw some examples similar to those on page 71 on the board or flip chart, and see whether the children can explain how to find the missing angles.

When the methods for calculating missing angles have been discussed, the children should work individually to complete both sheets. Because the main focus of this activity is reasoning about shapes and angle properties rather than calculation, you might allow some groups to use a calculator to find or check their answers. However, since these exercises do provide good practice in using mental addition and subtraction strategies, you may want to revise some of these at the start of the lesson or reflect on them during a plenary session.

The answers for page 70 are: a = 40°, b = 125°, c = 107°, d = 142°, e = 60°, f = 63°, g = 150°, h = 160°, j = 105°. The answers for page 71 are: a = 45°, b = 62°, c = 18°, d = 72°, e = 122°, f = 48°, g = 100°, h = 70°, j = 54°, k = 108°.

DISCUSSION QUESTIONS
- *How can you find that missing angle?*
- *What is a good way to check your answer?*
- *Which angles were easy to calculate? Which were harder?*

ASSESSMENT
Can the children calculate the missing angles? Can they explain their strategies for solving the problems?

VARIATION
The examples on pages 70 and 71 are drawn accurately so that, in every case, the children can check their answers by using a protractor. Note that using a protractor will give answers correct to within 1° of accuracy. The children could reflect on why, for these exercises, finding answers by calculation is more accurate than using direct measurement.

EXTENSIONS
- Create 'missing angle' problems using pentagons (the total of the internal angles is 540°).
- Create 'missing angle' problems for hexagons (the total is 720°).
- Consider what happens to the total of the angles as the number of sides increases.What would the total be for a 12-sided shape?

DEVELOPING SHAPE, SPACE & MEASURES

ANGLES QUIZ

A. State whether these angles are acute, right, obtuse, straight line or reflex.

1. 125° _____   2. 90° _____   3. 200° _____

4. 88° _____   5. 270° _____   6. 180° _____

7. 91° _____   8. 305° _____   9. 100° _____

B. Rewrite these sets of angles in order of size (smallest first):

1.   a right angle   200°   89°   a half turn

_____

2.   75°   a quarter turn   a reflex angle   180°

_____

3.   a straight line   90°   an acute angle   175°

_____

 Check your answers with a friend.

SEE 'DIFFERENT ANGLES', PAGE 59.

WHAT'S THE ANGLE? (1)

■ Estimate and measure each of these angles.

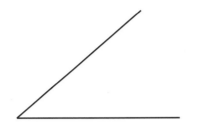

1. Estimate: ☐ Actual size: ☐

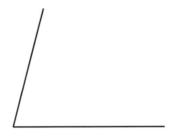

2. Estimate: ☐ Actual size: ☐

3. Estimate: ☐ Actual size: ☐

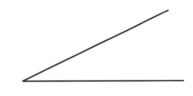

4. Estimate: ☐ Actual size: ☐

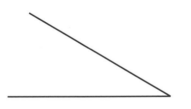

5. Estimate: ☐ Actual size: ☐

6. Estimate: ☐ Actual size: ☐

7. Estimate: ☐ Actual size: ☐

8. Estimate: ☐ Actual size: ☐

 Did your partner get the same answers?

SEE 'MEASURING ANGLES', PAGE 60.

NAME _____   DATE _____

WHAT'S THE ANGLE? (2)

■ Estimate and measure each of these acute and obtuse angles.

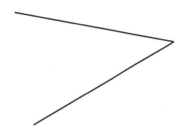

1. Estimate: ☐ Actual size: ☐   2. Estimate: ☐ Actual size: ☐

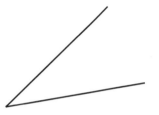

3. Estimate: ☐ Actual size: ☐   4. Estimate: ☐ Actual size: ☐

5. Estimate: ☐ Actual size: ☐   6. Estimate: ☐ Actual size: ☐

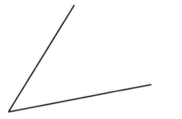
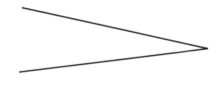

7. Estimate: ☐ Actual size: ☐   8. Estimate: ☐ Actual size: ☐

 Which angles were hardest to measure? What made them so tricky?

SEE 'MEASURING ANGLES', PAGE 60.

DRAW THE ANGLES

■ At the end of each line, draw a second line to show an angle at the point indicated. Mark each angle 'acute', 'right' or 'obtuse' as appropriate.

75°

120°

80°

130°

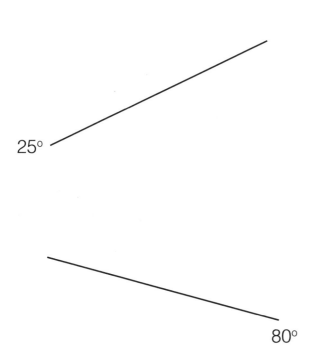

25°

80°

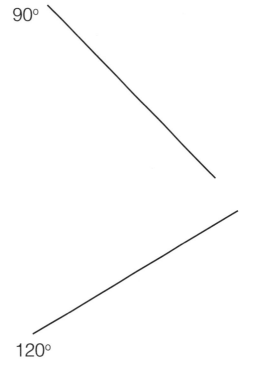

90°

120°

 Now get a friend to check your answers.

SEE 'DRAWING ANGLES', PAGE 61.

PHOTOCOPIABLE

NAME _____   DATE _____

ANGLES THAT MEET

■ Calculate the missing angles in these angle diagrams.

140° / a

55° \ b

73° \ c

d / 38°

70° \ e / 50°

81° \ f / 36°

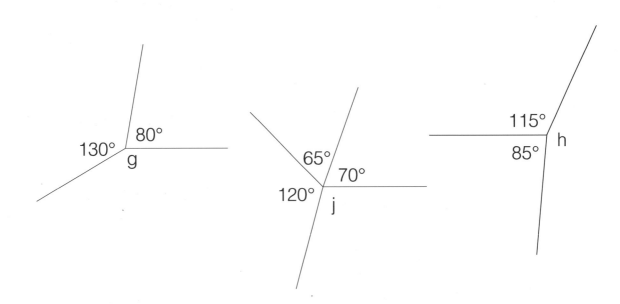

130° 80° g

65° 70° 120° j

115° 85° h

 Write down instructions on how to find the answers for a friend who has been away from school.

 How could you check your answers?

SEE 'ANGLE CHALLENGE', PAGE 64.

70

**DEVELOPING SHAPE,
SPACE & MEASURES**

NAME _____   DATE _____

ANGLES IN SHAPES

■ Calculate the missing angles in these shapes.

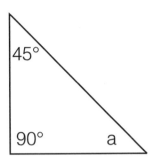

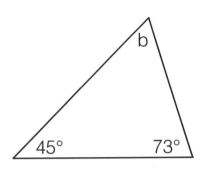

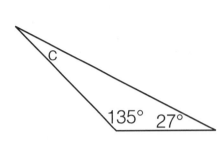

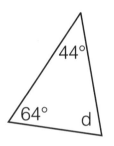

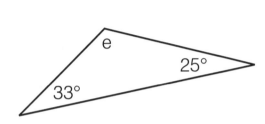

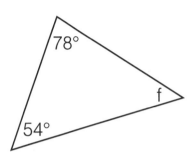

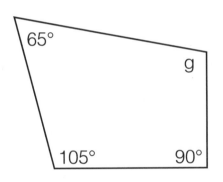

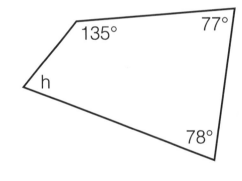

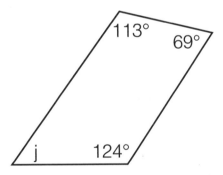

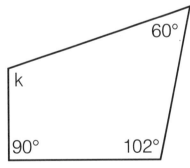

 Draw some more triangles and quadrilaterals, marking the angles accurately but leaving one angle out. Can your friend calculate the missing angle?

SEE 'ANGLE CHALLENGE', PAGE 64.

DEVELOPING SHAPE,
SPACE & MEASURES

MEASURING INSTRUMENTS

ruler

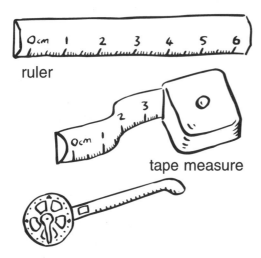
tape measure

trundle wheel

pan balance

weighing scales

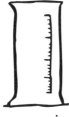
measuring cylinder

measuring jug

USEFUL WORDS

size	unit	scale
approximate	roughly	estimate
just more than	just less than	
length	distance	height
weight	mass	capacity
volume	area	perimeter
surface		

METRIC UNITS

- The unit of length is the **metre**.
- The unit of capacity is the **litre**.
- The unit of mass is the **gram**.

These units can be used with different **prefixes** for a larger or smaller scale:
- **milli-** means one thousandth
- **centi-** means one hundredth
- **kilo-** means one thousand.

Useful facts to remember

- 1000 metres (m) = 1 kilometre (km)
- 100 centimetres (cm) = 1 metre (m)
- 10 millimetres (mm) = 1 centimetre (cm)
- 1000 grams (g) = 1 kilogram (kg)
- 1000 millilitres (ml) = 1 litre (l)
- 100 centilitres (cl) = 1 litre (l)

Area is measured in m^2 or cm^2.
$10\,000cm^2 = 1m^2$

IMPERIAL UNITS

Imperial units are traditional (in Britain and America), but are being replaced by **metric units**, which are more mathematically simple to calculate with.

Common imperial units

- Length: inches, feet, yards, miles.
- Mass: ounces (oz), pounds (lb), stones.
- Capacity: pints, gallons.

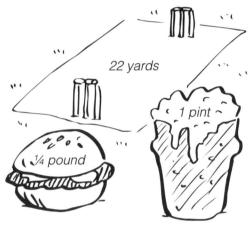

22 yards

1 pint

¼ pound

12 inches = 1 foot
16 ounces (oz) = 1 pound (lb)
14 pounds = 1 stone

Useful conversions

Metric to imperial		
1cm	=	0.39 inches
1 metre	=	1.09 yards
1km	=	0.62 miles
1 litre	=	1.76 pints
1kg	=	2.20lb
Imperial to metric		
1 inch	=	2.54cm
1 yard	=	0.91m
1 mile	=	1.61km
1 pint	=	0.57l
1 pound	=	0.45kg

UNITS OF TIME

60 seconds	=	1 minute
60 minutes	=	1 hour
24 hours	=	1 day
7 days	=	1 week
52 weeks / 365 days	=	1 year
1 leap year	=	366 days
10 years	=	1 decade
100 years	=	1 century
1000 years	=	1 millennium

How many days are there to the next millennium?

How many days of my life will I spend at school?

KEY IDEAS

- Consolidating the vocabulary of length and area.
- Choosing a suitable scale of (metric) units.
- Knowing relationships between metric and imperial units.
- Measuring the lengths of non-linear objects.
- Calculating areas.
- Comparing the areas of different oblong shapes.
- Exploring the relationship between area and perimeter.

The word 'length' covers a range of different attributes, particularly when we are concerned with 3-D objects: the height, depth and width of an object are all linear measures. It is important for the children to develop an understanding of these terms, particularly when they begin to measure objects such as cuboids, in order to distinguish between the different dimensions.

Although the children will have encountered the metric units of length (kilometres, metres and centimetres) before, at this stage they should be starting to appreciate the scale of these units and the mathematical relationships between them. They should also have opportunities to use millimetres. It is useful to reflect on the degree of accuracy needed in practical situations: if I'm travelling from London to Glasgow, a measurement to the nearest 10km is sufficient; but if I'm buying a pair of trousers, I need to be accurate to the nearest centimetre. Cross-curricular work, for example in geography and in design and technology, provides many useful opportunities for considering this theme. Imperial units also need to be considered, as well as the relationships between the metric and imperial systems.

The children need to appreciate the difference between measuring length and measuring area. Area is difficult to measure directly: the children should have moved beyond the counting of squares, and see area as the result of a calculation – for example, multiplying the length and width of a rectangle to give its area. However, we should try to avoid simply teaching a formula without paying attention to the underlying concepts, particularly when considering areas that are more difficult to calculate (such as 'L' shapes or triangles).

BY THE END OF Y5/P6, MOST CHILDREN SHOULD BE ABLE TO:

- use, read and write standard metric units of length and area, including their abbreviations and relationships
- use common imperial units for length, and know in particular that a mile is about 1600m
- suggest suitable units and measuring equipment to estimate or measure length (including height and distance)
- measure and draw lines to the nearest millimetre
- understand and use the formula for the area of a rectangle
- understand, measure and calculate the perimeters of rectangles and regular polygons.

BY THE END OF Y6/P7, MOST CHILDREN SHOULD BE ABLE TO:

- use, read and write standard metric units for length and area, including their abbreviations and relationships
- measure and record lengths and areas using decimals
- convert smaller to larger units and vice versa
- use common imperial units for length and know rough metric equivalents (such as: 8km is about 5 miles; a metre is about 3 feet and 3 inches)
- suggest suitable metric or imperial units and measuring equipment to estimate or measure length, and recognize appropriate units for a variety of situations
- calculate the perimeter and area of simple compound shapes that can be split into rectangles, and of right-angled triangles as half-rectangles
- start to use linear scales (such as 1cm to 1m or 1:10), particularly in geography or technology.

SOME COMMON MISCONCEPTIONS AND STRATEGIES FOR CORRECTING THEM

METRIC AND IMPERIAL

Children often confuse metric and imperial units of length. They should be encouraged to understand the historical relationship between the two systems, and to appreciate the mathematical and practical advantages of the metric system – particularly the conversion between different metric units, which are all based on relationships of 10, 100 or 1000. Such conversions are much easier to calculate than those involving imperial units, though they do require a good understanding of place value.

UNITS OF AREA

Children need to be constantly reminded that area is a two-dimensional attribute, and so is measured in cm^2 or m^2 rather than cm or m. Also, although 1m = 100cm, $1m^2 = 10\,000cm^2$ ($100cm \times 100cm$). Children often assume that since 1m = 100cm, $1m^2 = 100cm^2$.

A good way to demonstrate the fallacy of this is to lay out a square metre (use four metre sticks) and ask the children to fill the space with centimetre squared paper: they will soon realize that more than 100 centimetre squares are needed.

TWICE AS BIG

Children tend to assume that if a rectangle is made twice as long **and** twice as high, it will have twice the area. But since this is effectively two doublings, the area will be increased to four times its original value:

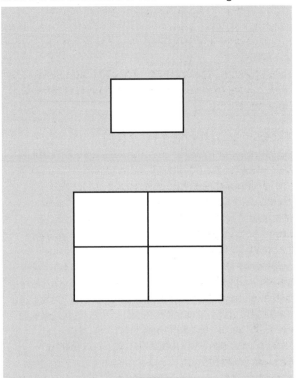

Help the children to understand this by showing the effects of doubling first the length, then the height, then both:

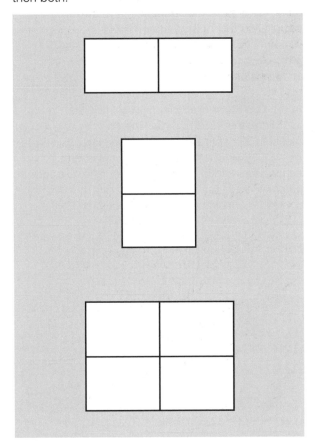

LENGTH & AREA

UNITS OF LENGTH

†† *Whole class, then pairs*
🕐 *50 minutes*

AIMS
To understand the relationship between different metric units of length. To measure lengths to the nearest millimetre.

WHAT YOU WILL NEED
Small objects to measure (such as pencils, rubbers and pencil cases), rulers, metre sticks.

WHAT TO DO
In a short warm-up session, lead the children to focus on the idea of how big different lengths are. A good way to do this is to ask them to visualize objects of given lengths. For example, can they imagine something that is 5cm/15cm/100cm/10m long? What about 1km or 1mm? Can they show you these lengths with their hands, or suggest suitable objects to represent the lengths?

Now remind the children of the relationship between different units of length: 100 centimetres are equal to a metre, 1000 metres make a kilometre. Ask whether anyone knows a unit smaller than a centimetre (a millimetre). Do they know how many of these there are in a centimetre? Also establish the relationship between millimetres and metres (analogous to that between millilitres and litres, with which the children should already be familiar).

Now ask for a volunteer and measure his or her height (alternatively, just say your own height) in metres. Ask: *How tall will that be in centimetres? In millimetres?* (For example, a height of 1.35m is equal to 135cm and 1350mm). Ask the children to measure a rubber carefully to the nearest millimetre, then to consider what the length would be in centimetres and in metres. (For example, a length of 42mm is equal to 4.2cm and 0.042m.) This is a good context to check the children's understanding of place value.

Now ask the children to draw up a table similar to the one shown below, make the various measurements and record them in millimetres, centimetres and metres. Decide which (and how many) objects they should measure.

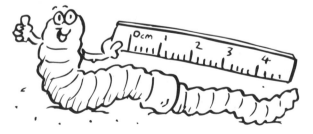

Object	millimetres	centimetres	metres
your height			
length of room			
width of a desk			
length of a book			
thickness of a dictionary			
pencil case			
width of a pencil sharpener			

Note that the initial choice of unit will vary: the children should make the initial measurement in the most appropriate unit before rewriting it in the other units. For example, they may choose to measure the length of the room in metres, the length of an exercise book in centimetres and the width of a pencil sharpener in millimetres; these should be recorded in the appropriate column before the corresponding amounts are written in the other two columns.

In a plenary session, discuss the children's results.

DISCUSSION QUESTIONS
● *What is the difference between a millimetre and a centimetre/a metre and a kilometre? What might each unit be used to measure?*
● *What is the best unit for measuring the length of the room/the length of your pencil case/the width of a pencil sharpener?*
● *When do you need to use millimetres? When would it not be necessary to measure to the nearest millimetre?*

ASSESSMENT
How accurately can the children measure with millimetres? Can they convert comfortably between metres, centimetres and millimetres, understanding the relationships between these units?

EXTENSIONS
● Ask questions that involve relating kilometres to other units. For example: *How tall are you in kilometres? How many centimetres from London to Birmingham?* The children can suggest similar problems.
● Photocopiable page 86 provides further individual practice in the relationships between different metric measures of length. The answers are: A 1) 460cm 2) 70cm 3) 7000cm; B 1) 4.53m 2) 1.6m 3) 0.28m; C 1) 87mm 2) 50mm 3) 580mm; D 1) 3.8cm 2) 1.1cm 3) 0.7cm; E 1) 5km 2) 3.03km 3) 0.8km; F 1) 1600m 2) 5070m 3) 50 200m; G 1) 5m 2) 2.096m 3) 0.467m; H 1) 1674mm 2) 760mm 3) 5200mm.
● Photocopiable page 87 covers similar ground, but in a problem-solving context. The questions can be used for further whole-class discussion on this topic. The answers are: 1) 0.8m 2) 22cm 3) 230 metres 4) 1.2m 5) 12 000m 6) The same length.

IMPERIAL LENGTHS

†† *Whole class, then individuals*

🕐 *50 minutes*

AIMS
To become familiar with imperial measures of length. To convert between metric and imperial measures.

WHAT YOU WILL NEED
Rulers, pencils, calculators, photocopiable page 88, graph paper.

WHAT TO DO
Explain to the children that the imperial system of measurement is traditional in Britain and America, and is still widely used (for example, on road signs) even though the metric system has become standard in science and industry. There are four main imperial units of length: inches, feet, yards and miles. You could look at rulers (many still have 12 inches/1 foot marked on one side) or an old metre stick with 1 yard marked on one side. The children can use such rulers to make a direct comparison between an inch and a centimetre. Explain the relationships between units in the imperial system: 12 inches = 1 foot, 3 feet = 1 yard and 1760 yards = 1 mile. These relationships can be used for a quick mental warm-up:

- How many inches in 5 feet?
- How many feet in 8 yards?
- How many inches in 3 yards?

It's also interesting to consider how many inches there are in a mile (1760 × 12 × 3), and reflect on how much simpler the units of length are in the metric system (see 'Units of length', page 76).

Give each child a copy of photocopiable page 88. In the first and second sections, they need to use the information provided to make simple conversions. In order to avoid confusion, the full names are used for the imperial units. The conversions can be worked out with a calculator, or used for practice in multiplying decimals on paper. You may prefer the children to round their answers to the nearest whole number. In the third section, the children need to convert units from metric to imperial or vice versa in order to make a comparison.

In a plenary session, review the answers and check any that the children found difficult. Converting metres to yards (or vice versa) is sometimes confusing, because the numbers are quite similar. The answers are: 1. a) 1.56 inches, b) 8.72 yards, c) 3.1 miles, d) 4.68 inches, e) 10.54 miles, f) 16.35 yards, g) 146.32 miles, h) 2.457 inches, j) 20.819 yards. 2. a) 20.32cm, b) 5.46m, c) 4.83km, d) 18.2m, e) 8.89cm, f) 177.1km, g) 12.7cm, h) 33.005km, j) 136.5m. 3. a) 11 inches, b) 13 miles, c) 22 metres, d) 6 inches, e) 250 miles.

DISCUSSION QUESTIONS
- *What units of length do you know?*
- *What are the advantages/disadvantages of each system?*
- *Which units are easiest to understand, the metric or imperial?*

ASSESSMENT
Can the children convert lengths between the metric and imperial systems? Do they appreciate the mathematical differences between the two systems of measures?

EXTENSIONS
- The activity 'Imperial mass and capacity' (page 95) explores the relationships between metric and imperial units for those measures. You may prefer to consider that activity as an extension of this work.
- The children can create simple line graphs for converting between metric and imperial measures: kilometres and miles, metres and yards or centimetres and inches. They could use these to check their answers to part 3 of page 88. It is useful to consider why these graphs will not give answers as accurate as those obtained from the numerical calculations (though in many 'real life' situations, this level of accuracy would be quite acceptable).
- If they have completed the activity 'A giant handprint' (page 85), the children can convert all of the Giant's measurements into imperial lengths. Similarly, they can extend the table of measures from the 'Units of length' activity (page 76) to include a column for imperial equivalents.

LENGTH & AREA

RECTANGLE AREAS AND PERIMETERS

†† Whole class, then individuals
🕐 50 minutes

AIMS

To revise finding the area and perimeter of a rectangle, extending to 'awkward' lengths. To recognize that area and perimeter have different units.

WHAT YOU WILL NEED

Pencils, rulers, 1cm squared paper, photocopiable page 89.

WHAT TO DO

Ask the children to draw a rectangle 4cm × 3cm, using 1cm squared paper. Now ask them to count the number of squares inside the rectangle. Check that they understand that this represents the area of the rectangle. Instead of counting squares, do they know a better way to work out the area? They should recall that the area can be found by multiplying the length and the width. You might like to confirm this by working through a few more examples.

Now establish the perimeter. Again, this can be done by measuring; but the children should recall that it is easier to add the length and the width, then multiply by 2. At a basic level, it is sufficient that they can explain how to find the perimeter without measuring all the way round; however, older or more confident pupils may start to express the relationship using an abbreviated formula.

Ask the children to draw two more rectangles (using their own choice of measurements), and to find the area and perimeter of each. Remind them of the need to distinguish between units for area (cm²) and units for perimeter (cm).

Now draw a rectangle on the board, marking its length as 93cm and its width as 27cm. Point out that for such a large rectangle, counting squares or measuring each side in turn would be a very slow method – hence the value of having a method of calculation. Calculate the area and perimeter together.

Give each child a copy of photocopiable page 89. The children can use a calculator either to carry out or to check the calculations, as you prefer.

Go through the answers in a plenary session:
1. area 33cm² perimeter 28cm, 2. area 36cm² perimeter 26cm, 3. area 76cm² perimeter 46cm, 4. area 360cm² perimeter 92cm, 5. area 1350cm² perimeter 210cm, 6. area 49.8cm² perimeter 28.6cm, 7. area 108cm² perimeter 50.8cm.

DISCUSSION QUESTIONS
● *What is a better way to find a rectangle's area?*
● *What is a better way to find a rectangle's perimeter?*
● *Why do we use cm² for area and cm for perimeter?*

ASSESSMENT

Can the children distinguish between area and perimeter? Can they explain in words how to calculate each? Can they calculate the area and perimeter of a rectangle accurately, using appropriate units?

EXTENSIONS
● The activity 'Rectangle 24' (page 81) investigates the relationship between perimeter and area in a more open-ended way.
● More confident children could consider how to find the length of a rectangle when you know its width and either its area or its perimeter.

AREA OF A TRIANGLE

†† Whole class, then individuals
🕐 50 minutes

AIM

To demonstrate practically how the area of a triangle relates to that of a rectangle.

WHAT YOU WILL NEED

Pencils, rulers, scissors, plain paper.

WHAT TO DO

Ask the children to draw a rectangle measuring, say, 8cm × 6cm. Confirm that they know its area. Now ask them to draw a diagonal line:

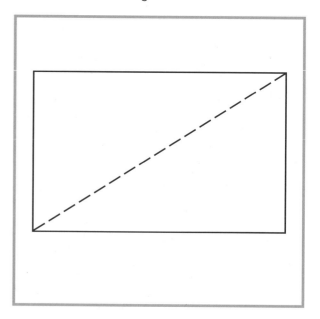

Ask: *What will happen when the rectangle is cut along this line? What can you say about the area of each piece?* If the children cut carefully along the diagonal, they should find that one piece fits exactly over the other. Each triangle therefore has an area of 24cm². Ask them to draw another rectangle (they can choose the size) and decide what the area will be of each triangle made by cutting along its diagonal.

Ask the children whether they notice anything else about the triangles they have cut out. They should see that the triangles are all right-angled. Ask them whether other, non-right-angled triangles cut from the rectangle will have half the area, or whether it is only true for right-angled ones. Again, they should draw a rectangle (starting with 8cm × 6cm), and draw two lines, starting from different corners along one side of the rectangle and meeting anywhere along the opposite side. Encourage pairs of children to create different triangles (see figure below).

Ask the children to predict the area of the larger triangle. To see that this triangle is half of the rectangle, they should cut along the pencil lines and flip the two smaller triangles over – together, these triangles should cover the whole area of the larger triangle. Let the children test this result by experimenting with a different rectangle.

When the children are comfortable with this, ask them to draw a triangle of any kind. Now ask them to find its area. Explain that they can do this by imagining or drawing a rectangle enclosing the triangle; lead them to realize that they can find the area of the triangle by halving the area of this rectangle. (The dimensions of the rectangle are equivalent to the base and height of the rectangle. It is from this that we derive the formula that the area of a triangle = ½ × base × height.)

They should then write an account (perhaps using the writing frame on page 128), explaining in their own words what they have found out about the area of a triangle.

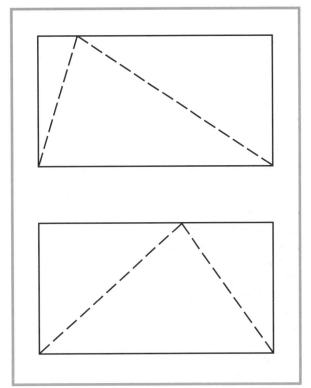

DISCUSSION QUESTIONS
● *What shapes will you make when you cut across the diagonal of the rectangle?*
● *How does the area of the triangle compare to that of the rectangle?*
● *What is a good way to find the area of any triangle?*

ASSESSMENT
Can the children understand and explain in their own words how to find the area of a triangle?

EXTENSIONS
● The children can draw the rectangles on 1cm squared paper and establish that the area of the triangle is half that of the rectangle by counting squares.
● More confident children may be ready to tackle calculations based on the formula for the area of a triangle: area = ½ × base × height.

LENGTH & AREA

SORT THE MEASURES

👥 *Whole class, then groups*
🕐 *50 minutes*

AIM
To revise the metric units of measurement for length, mass and capacity.

WHAT YOU WILL NEED
Sugar paper (one large sheet per group of three or four), pens, a variety of devices for measuring length.

WHAT TO DO
This activity can be used when beginning any unit of work on measure: it has been placed here, but could be used just as effectively at the start of work on mass and capacity. It is essentially a brainstorming session that will enable you to assess informally which units the children are familiar with, and which they may be less sure of.

Ask the children to tell you all the metric units (for length, mass and capacity) that they know. Scribe these on the board or flipchart. A reasonable selection might include metres, centimetres, millimetres, kilometres, grams, kilograms, litres and millilitres. It is likely that they will also mention miles, pounds or pints (imperial units); but for present purposes, concentrate on metric units.

Now ask the children to work in groups of three or four. Each group should discuss the list of units to decide:
● what each unit is used to measure
● the size of the unit (by drawing an object that it can be used to measure)
● how some of the units in any one category (length, mass or capacity) are related to each other.

Give each group a large sheet of sugar paper and ask them to produce a poster giving information about the units. Each group should select several units. The 'Measures glossary' (pages 72–73) can be used to support groups if necessary.

In a plenary session, discuss the content of the children's posters and address any discrepancies (for example, regarding the relationships between units). The posters can be displayed to support further work on this topic.

DISCUSSION QUESTIONS
● *What are metres/grams/litres used to measure?*
● *How big is a... [any particular unit]?*

ASSESSMENT
What units are the children familiar with? Can they identify the appropriate unit for measuring a particular item? Are they sure of the numerical relationships between different units?

EXTENSIONS
● This activity could include units of area if the children are already familiar with these.
● You could have a similar brainstorm for imperial units, or include imperial units in the activity and make the sorting of the units into imperial and metric one of the features for the children to include in their posters.

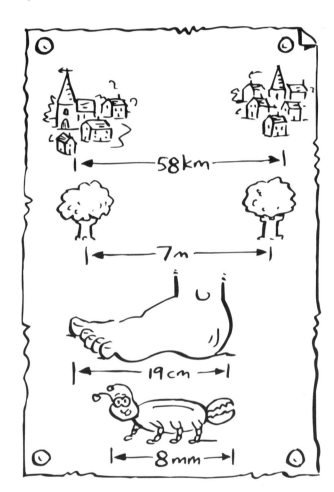

MEASURING CIRCLES

👥 *Whole class, then pairs*
🕐 *50 minutes*

AIM
To measure the diameter and circumference of a circle in a practical context.

WHAT YOU WILL NEED
A collection of cylinders, tubes and other 3-D objects with circular faces, plastic circles, compasses, rulers, string, tape measures, paper, pencils.

WHAT TO DO

Start by explaining the following circle vocabulary:
● **Diameter** – the distance all the way across a circle through the centre.
● **Radius** – the distance from the edge to the centre of a circle (ie half the diameter).
● **Circumference** – the perimeter of a circle.
Show the children some objects with circular faces. Ask them how they might measure the diameter and circumference of each circle. A ruler is clearly not suitable for measuring the circumference – so demonstrate how to place a length of string around the circumference, mark it, pull it straight, then measure it with a ruler. For large cylindrical objects, a tape measure can be used.

Distribute a range of circular and cylindrical objects to each table. The children can also draw circles with compasses. Each pair should aim to measure the circumference and diameter of around eight different circles, recording their results in a table (see below).

Bring the class back together and compile a class table, taking one or two results from each group. Discuss what the children notice about the relative sizes of the circumference and diameter; they should notice that the circumference is always just over three times the diameter. Depending on the age or experience of the class you may want to discuss pi or π (the ratio of the circumference to the diameter, approximately 3.14) more formally.

Object	Circumference	Diameter
Baked bean tin		
Red plastic hoop		

DISCUSSION QUESTIONS

● *What would be the best way to measure the circumference of this object? How about the diameter?*
● *What do you notice if you compare the circumference to the diameter?*
● *How would the circumference compare to the radius?*

ASSESSMENT

Can the children make accurate measurements of the circumference and diameter of different objects, selecting from a range of equipment?

EXTENSIONS

● The children can add a fourth column to the table and record the ratio of circumference to diameter for each object.

● They can plot a graph of diameter (on the x-axis) against circumference (on the y-axis) for several circles. Because of the constant ratio, they should find that these points all fall in a straight line.

RECTANGLE 24

†† *Whole class, then pairs*
🕐 *50 minutes*

AIM

To explore the relationship between the area and the perimeter of a rectangle.

WHAT YOU WILL NEED

1cm squared paper, rulers, pencils.

WHAT TO DO

Give out one sheet of 1cm squared paper per pair. Ask the children to find and draw as many rectangles as they can with a perimeter of 24cm. After 10 minutes, ask them to report back. Collect their findings together: they should have found rectangles 11cm × 1cm, 10cm × 2cm, 9cm × 3cm, 8cm × 4cm, 7cm × 5cm and 6cm × 6cm. Now ask the children to predict which of these has the largest area. Ask: *What is special about this shape?* [It is a square: its sides are all the same.]

Now ask the children to investigate rectangles with a perimeter of 20cm or 32cm in the same way. Ask them to predict which of the solutions will have the largest area. Once it has been established that in each case, the rectangle with the largest area is a square, ask them to investigate rectangles with a perimeter of 26cm. They may conclude that the rectangle with the largest area is 7cm × 6cm, which has an area of 42cm². However, what if you did not stick to whole centimetres? Ask them to check a 6.5cm × 6.5cm rectangle – this has a perimeter of 26cm and an area of 42.25cm².

**DEVELOPING SHAPE,
SPACE & MEASURES**

LENGTH & AREA

Try some more examples, using different perimeters. Older or more able children should realize that, for any perimeter, the rectangle with the largest area can be found by simply dividing the perimeter by 4: this will give the side length of a square.

DISCUSSION QUESTIONS
● *How many different rectangles have you found with a perimeter of 24cm?*
● *What kind of rectangle has the largest area for any perimeter?*
● *Why is this? ['The space is clumped together' or 'The shape is not stretched at all.']*

ASSESSMENT
Can children find a range of rectangles for a given perimeter? Can they see a pattern in their results (for any given perimeter, the long thin rectangles have a smaller area than the more nearly square ones)?

VARIATION
● The children can start with a fixed area (for example 24cm²) and find as many different rectangles as they can with that area. *Which has the largest perimeter? Which has the smallest?*

EXTENSIONS
● The children can use the writing frame on page 128 to reflect on what they have found out about area and perimeter.
● Photocopiable page 91 is useful for individual consolidation of this topic. Alternatively, use the questions it poses for further whole-class discussion. The answers are: A. 3cm × 2cm, B. 3cm × 3cm, C. 7cm × 2cm, D. 10cm × 4cm, E. 6cm × 1cm, F. 8cm × 2cm.
● The children can investigate pentominoes or hexominoes (see resource pages 122 and 123). *Each of the figures in this set has the same area, but which has the largest perimeter? Which has the smallest?* With the hexominoes, the children could try to create further hexominoes with a larger or a smaller perimeter than the ones on the sheet. They could go on to investigate heptominoes (shapes made up of seven squares).

SPLITTING SHAPES

†† *Whole class, then individuals*
🕘 *50 minutes*

AIM
To find the areas of compound shapes.

WHAT YOU WILL NEED
1cm squared paper, pencils, rulers, photocopiable page 90.

WHAT TO DO
Draw the following diagrams on the board or flip chart (or an OHT), and ask the children to copy them carefully onto squared paper.

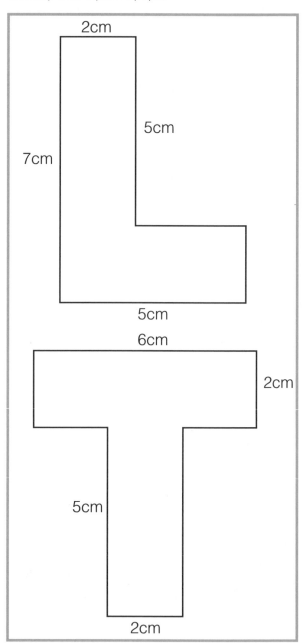

Now ask them to discuss with a partner how they could find the area of each figure. Clearly, they could do it by counting squares – but remind them how, when they considered the areas of rectangles, they found a more efficient method of calculation (see 'Rectangle areas and perimeters', page 78). If no one suggests it, prompt the children to try splitting either shape into rectangles and calculating the area of each piece, then adding the parts together to find the whole. Encourage them to see that for each figure, there is more than one way of doing this – indeed, it is useful to use an alternative way of dividing the shape as a check.

Once the area of each figure has been agreed, distribute copies of photocopiable page 90 and let the children use these for individual practice. Support those children who need help. Follow up with a plenary session focusing on the different ways that the children have found to split the shapes. The answers are: A. 198cm^2 B. 56cm^2 C. 112cm^2 D. 684cm^2.

DISCUSSION QUESTIONS
- *How can we split this shape into rectangles?*
- *Can anyone see a different way to do it?*

ASSESSMENT
Can the children see how to split compound shapes into rectangles? Can they use this method to calculate the areas of the compound shapes? Can they check by splitting the shape a different way?

EXTENSION
The children can create more complex compound shapes and challenge their friends to find the areas.

ON THE SURFACE

†† *Whole class, then groups*
🕐 *50 minutes*

AIM
To calculate the surface area of a cuboid.

WHAT YOU WILL NEED
A collection of cuboid boxes with widely varying dimensions (such as cereal boxes, tea-bag boxes, chocolate boxes and tin foil boxes), rulers, paper, pencils, calculators.

WHAT TO DO
Confirm that the children know that a cuboid has six surfaces, and that each of these is a rectangle. Show them a cuboid box and ask them what they can say about the sizes of the six rectangles. They should notice that opposite sides of the cuboid have the same size. Explain that the total area of all six surfaces of a cuboid box is called its **surface area**.

Ask the children to find, by measuring and calculating, the surface area of several different cuboid boxes. Hold up several different boxes and see whether they can predict which has the greatest surface area. Children often predict a long thin box as having the largest surface area and are surprised that a shorter, wider box has more. Opening the boxes up and comparing how much card is used can help. Consider the accuracy of measurement that is needed; the children can measure to the nearest centimetre or millimetre.

Give several boxes to each group. They should sketch each box and record its dimensions, then calculate the surface area by finding the area of each rectangular face and adding them together.

Conclude by reviewing the children's results. Ask them how they could find the total surface area of any cuboid. Encourage them to realize that, since the rectangles of a cuboid form equal pairs, they can find the total surface area by finding the total area of the three different rectangles and then doubling that.

DISCUSSION QUESTIONS
- *What do you notice about opposite faces of the cuboid?*
- *What measurements will you need to make?*
- *How do you check you have included all six faces?*

ASSESSMENT
Can the children make accurate measurements and calculations in order to establish the surface area of a cuboid? Can they explain, in their own words, how they would find the surface area of a cuboid?

DEVELOPING SHAPE, SPACE & MEASURES

EXTENSION

This activity can be linked to work on the volume of a cuboid (see the section on 'Mass and capacity', page 92). Children often confuse the calculations for volume and surface area.

DOUBLE UP

†† *Whole class, then groups*
🕐 *50 minutes*

AIM

To investigate the effect of doubling the dimensions of a rectangle on its area.

WHAT YOU WILL NEED

Squared paper, pencils, rulers, resource page 128.

WHAT TO DO

Ask the children to draw a rectangle 5cm × 4cm and to record its length, width, perimeter and area. Now ask them what will happen if they draw a rectangle double the dimensions – that is, double the length **and** double the width: can they predict the perimeter and area of the new rectangle? They are likely to predict that both the perimeter and the area will have doubled.

Now ask them to draw the new shape and check their predictions. Can they explain what has happened? (The perimeter has doubled, but the area has quadrupled.) In case they suspect that this result was a fluke, ask them to check with another rectangle: they can choose the dimensions.

Once it has been agreed that the same thing happens for any rectangle, ask the children to form small groups and discuss why this happens. Now ask the groups to go back to the 5cm × 4cm rectangle and to try to find a way of altering its dimensions so that the area is doubled. (The solution

to this problem is to double only one of the dimensions: either a 10cm × 4cm or a 5cm × 8cm rectangle will have double the area of the original shape.)

The children can conclude the lesson by writing an account of their findings, using the writing frame on resource page 128.

DISCUSSION QUESTIONS

● *What will happen if you double the length **and** the width of the rectangle?*
● *How can you explain what has happened to the area?*
● *How can we double the area of any rectangle?*

ASSESSMENT

Can the children explain, in their own words, what happens when the dimensions of a rectangle are doubled? Can they explain why this happens? Can they explain how to double the area of any rectangle?

EXTENSION

For a similar investigation of capacity or volume which can be used as an extension of this activity, see 'Which holds the most?' (page 101).

A GIANT HANDPRINT

†† *Whole class, then groups*
⏱ *50 minutes*

AIMS
To measure length in a practical context. To solve a problem by using a ratio.

WHAT YOU WILL NEED
Rulers, tape measures, calculators, pencils, paper, a giant handprint. (Draw around an adult hand, shading in a few veins for effect, then enlarge it several times – as large as you can. Alternatively, draw a very large 'handprint' freehand on a sheet of sugar paper.)

WHAT TO DO
Display the handprint. Tell the children that the caretaker has found this handprint on the school wall, just outside the headteacher's office. You could tie this in with a suitable story such as 'The Selfish Giant' by Oscar Wilde, which is available in several editions.

The first task for the children is to consider how tall the giant must be. The place where the handprint was found does not provide an answer. Let them discuss this problem in groups before talking it through with the whole class. Ways to solve the problem include:

● calculating the ratio of a child's height to his or her hand size, then multiplying the giant's hand size by this ratio
● finding out how many times larger the giant's hand is than a child's, then multiplying the child's height by that ratio.

Both of these methods assume a constant ratio – that is, they assume that the giant's body has the same proportions as a child's. The children may realize that the ratio used will depend upon the child. To check this, they can all measure their height and handspan and calculate the ratio; they may conclude that taking an average is a good idea.

Once the giant's likely height has been established, set the groups different problems from the following list (or let them choose something similar to investigate):

● Could the giant get into the headteacher's office?
● Could he walk along the hall?
● What size cutlery does the giant use?
● How long are the giant's arms?
● What size furniture would the giant need?
● What size pens, rubbers and pencil sharpeners does the giant use?

All of these require the children to undertake measuring in the classroom or around the school.

Bring the class back together to discuss their findings. Their records could form a display about the giant, alongside pictures or models of the giant's belongings (such as the cutlery).

DISCUSSION QUESTIONS
● How can you compare your height to your handspan? [You can introduce the word 'ratio' here if the children are not already using it.]
● How can we find out the giant's height?
● Is there another way to check?
● How accurate will our answers be? Why?

ASSESSMENT
Can the children find a way to solve each problem, using ratios? Can they identify the necessary measurements and make them accurately?

EXTENSIONS
In a subsequent session, you might consider other aspects of the giant that involve measures other than length. Different groups of children could research different problems, such as:
● What does the giant weigh?
● What does the giant's new-born baby weigh?
● How much would the giant eat in one day?
● How much would the giant drink in one day?
● How many millilitres will the giant's glass hold?

85

NAME _____   DATE _____

METRIC LENGTHS

Remember: 1km = 1000m, 1m = 100cm, 1cm = 10mm.

A. Rewrite these lengths in cm.

    1) 4.6m _____   2) 0.7m _____   3) 70m _____

B. Rewrite these lengths in metres.

    1) 453cm _____   2) 160cm _____   3) 28cm _____

C. Rewrite these lengths in mm.

    1) 8.7cm _____   2) 5cm _____   3) 58cm _____

D. Rewrite these lengths in cm.

    1) 38mm _____   2) 11mm _____   3) 7mm _____

E. Rewrite these lengths in km.

    1) 5000m _____   2) 3030m _____   3) 800m _____

F. Rewrite these lengths in metres.

    1) 1.6km _____   2) 5.07km _____   3) 50.2km _____

G. Rewrite these lengths in metres.

    1) 5000mm _____   2) 2096mm _____   3) 467mm _____

H. Rewrite these lengths in mm.

    1) 1.674m _____   2) 0.76m _____   3) 5.2m _____

 Which lengths are easiest to convert? Which are confusing?

SEE 'UNITS OF LENGTH', PAGE 76.

NAME   DATE

WHICH IS GREATER?

■ For each of these pairs, write down which is greater. Show how you worked it out.

1. A strip of liquorice that is 0.8m long or one that is 85mm long? 	2. A book that is 22cm wide or one that is 0.2m wide? 
3. A distance of 230 metres or a distance of 20 000cm? 	4. A snake that is 1.2m long or one that is 1100mm long? 
5. A town that is 11km away or one that is 12 000m away?	6. A corridor that is 30 metres long or one that is 30 000mm long?

 Now make up some more puzzles like these for your friends to try. Make sure you know the correct answers before you swap problems.

SEE 'UNITS OF LENGTH', PAGE 76.

LENGTH & AREA

NAME DATE

CONVERTING LENGTHS

1cm = 0.39 inches	1 inch = 2.54cm
1m = 1.09 yards	1 yard = 0.91m
1km = 0.62 miles	1 mile = 1.61km

■ Use the information in the box above to make the following conversions:

1. Metric to imperial
   a) 4cm = __________
   b) 8m = __________
   c) 5km = __________
   d) 12cm = __________
   e) 17km = __________
   f) 15m = __________
   g) 236km = __________
   h) 6.3cm = __________
   j) 19.1m = __________

2. Imperial to metric
   a) 8 inches = __________
   b) 6 yards = __________
   c) 3 miles = __________
   d) 20 yards = __________
   e) 3.5 inches = __________
   f) 110 miles = __________
   g) 5 inches = __________
   h) 20.5 miles = __________
   j) 150 yards = __________

3. In each case, which is the longer?
   a) 11 inches or 27 centimetres? __________
   b) 20km or 13 miles? __________
   c) 24 yards or 22 metres? __________
   d) 15 centimetres or 6 inches? __________
   e) 250 miles or 400km? __________

On the back of the sheet write some instructions for a friend, explaining how to solve these problems.

SEE 'IMPERIAL LENGTHS', PAGE 77.

RECTANGLE CHECK-UP

■ Measure these two rectangles and find their area and perimeter.

1. Area: _____   Perimeter: _____

2. Area: _____

Perimeter: _____

■ Complete this table by working out the area and perimeter of each rectangle.

	Length	Width	Area	Perimeter
3.	19cm	4cm		
4.	36cm	10cm		
5.	90cm	15cm		
6.	8.3cm	6cm		
7.	20cm	5.4cm		

 On the back of this sheet, explain in your own words how to find the perimeter and the area of any rectangle.

SEE 'RECTANGLE AREAS AND PERIMETERS', PAGE 78.

LENGTH & AREA

NAME

DATE

SHAPE SPLITS

■ Find the area of each of these shapes. They are not drawn to full size.

A

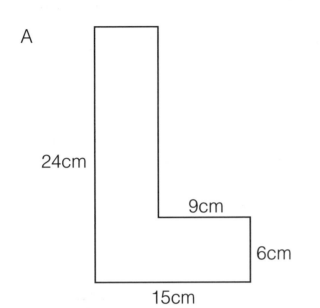

24cm

9cm

6cm

15cm

B

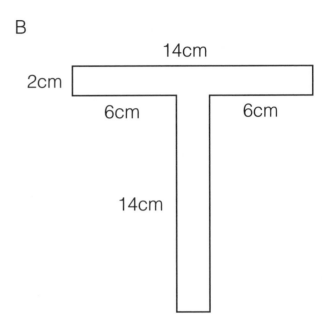

14cm

2cm

6cm   6cm

14cm

C

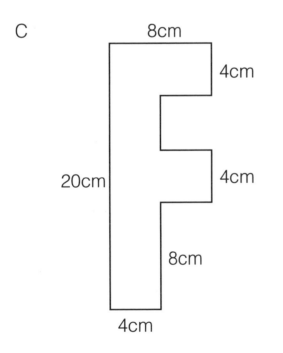

8cm

4cm

20cm

4cm

8cm

4cm

D

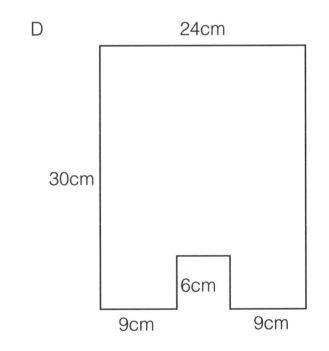

24cm

30cm

6cm

9cm   9cm

 Can you find a different way to work out each of these areas?

SEE 'SPLITTING SHAPES', PAGE 82.

FIND THE RECTANGLE

■ Find a rectangle that will match each description. Draw the rectangles to confirm the solution, using a different colour for each one. Mark the length and width of each rectangle. You may need some extra paper for calculations.

A. Perimeter 10cm, area 6cm^2.

B. Perimeter 12cm, area 9cm^2.

C. Perimeter 18cm, area 14cm^2.

D. Perimeter 28cm, area 40cm^2.

E. Perimeter 14cm, area 6cm^2.

F. Perimeter 20cm, area 16cm^2.

 On the back of the sheet, draw some more rectangles. Write down the area and perimeter for each one.

 Can your partner work out the length and width of each of your rectangles without measuring?

SEE 'RECTANGLE 24', PAGE 81.

DEVELOPING SHAPE,
SPACE & MEASURES

KEY IDEAS

- Reading a variety of scales and approximating sensibly in a range of practical situations.
- Using a full range of metric units.
- Understanding relationships between metric and imperial units.
- Understanding the relationship between capacity and volume.

There are three main aspects to the work presented in this chapter:

1. Consolidation of the use of a range of measuring equipment. Weighing scales and measuring jugs and cylinders come in a bewildering range of shapes and sizes, and children should be given opportunities to choose the most appropriate equipment to use in different situations.

2. Children need to understand the full range of units, as well as the relationships between them. Whereas '1 metre' can be seen, and comparing it to a child's height gives an immediate sense of its size, children need more help to appreciate the size of '1 litre' or '1 kilogram'. The relationships between scales of units in the metric system are all based on relationships of 10, 100 or 1000, but the system of prefixes is not as readily linked to a sense of relative size as, say, gallons and pints or stones and ounces.

Another interesting aspect of the relationships between units in the metric system is the interdependence of different measures. For example, 1ml is defined as the amount of water that takes up $1cm^3$ of space and weighs 1g; 1 litre of water weighs 1kg. Such simple and elegant relationships do not exist in the imperial system, where each type of unit was developed independently.

3. Although the children need practice in using the relationships between units to give them confidence in the numerical aspects, a range of practical activities are also needed to give them a proper appreciation of the uses of different measures. Some of the activities in this chapter can be linked directly to aspects of the primary science curriculum, and it is through work in other curriculum areas that the children will really be using many of these ideas.

BY THE END OF Y5/P6, MOST CHILDREN SHOULD BE ABLE TO:

- use, read and write standard metric units for mass and capacity, including their abbreviations and relationships
- know the commonly used imperial units for mass and capacity
- suggest suitable units and measuring equipment to estimate or measure mass and capacity
- record estimates and readings from scales to a suitable degree of accuracy.

BY THE END OF Y6/P7, MOST CHILDREN SHOULD BE ABLE TO:

- use, read and write standard metric units, including their abbreviations and relationships
- convert smaller to larger units and vice versa
- use imperial units for mass and capacity, and know the rough metric equivalents
- suggest suitable units (metric or imperial) and measuring equipment to estimate or measure mass and capacity
- record estimates and readings from scales to a suitable degree of accuracy
- understand that volume is conserved when the shape of a container is changed.

COMMON MISCONCEPTIONS AND STRATEGIES FOR CORRECTING THEM
RELATIVE SIZES OF DIFFERENT UNITS

Children need to have a sense of what might be measured in grams rather than kilograms, or litres rather than millilitres. For example, a packet of sweets will be measured in grams, while a sack of potatoes will be in kilograms.

Examining the labels of food and other packages is useful for building up a sense of units.

MASS AND WEIGHT

The difference between 'mass' and 'weight' needs to be linked to science work on forces. Weight is a force, measured in newtons (N). The children need to appreciate that weight is actually the pull of gravity on a mass. We would 'weigh' less on the Moon, where the force of gravity is less powerful, but our 'mass' (amount of substance) would remain the same.

What makes this difficult to appreciate is that for most practical purposes, and in the activities here, mass and weight are measured in the same way.

Linguistically, this is further complicated by the fact that we talk about 'weighing objects' (to find their mass), say that groceries are sold 'by weight', worry about our 'weight', and so on.

Talk about 'finding the mass of' something rather than 'weighing' it. Show the children scales and balances that are used to find mass and newton meters that are used to find a pulling force (or weight). Point out to the children that when they stand on the bathroom scales, they are really measuring the pushing force exerted by their mass – but the scale is corrected to show their mass (kg) rather than their weight (N).

The distinction between mass and weight does not need to be known until KS3 in England and Northern Ireland, but is required at KS2 in Wales – and besides, it is important for all children not to have to 'unlearn' incorrect terms as they go along. In Scotland, gravity should be introduced by P6 and newtons between P7 and S2.

CAPACITY AND VOLUME

Children often confuse the ideas of volume and capacity. The volume of an object is how much space it takes up. Solid volume is usually measured in cm^3, while liquid volume is measured in ml. The equivalence of these units can be used to find the volume of an irregular object by water displacement (as in the famous Archimedes problem). Capacity refers to the maximum volume (usually of a liquid) that a container can hold. (Note that the Scottish *5–14 Guidelines for Mathematics* refer to 'volume' rather than 'capacity'.)

A useful example to illustrate the distinction is a Thermos flask, which has a thick lining. Its volume (how much space it takes up) is much greater than its capacity (how much it can hold inside).

MASS & CAPACITY

A SHEET OF PAPER

†† *Whole class, then pairs*
🕒 *50 minutes*

AIM
To calculate small masses and volumes by grouping items.

WHAT YOU WILL NEED
Sheets of A4 paper (at least 100), several packs of playing cards.

WHAT TO DO
Hold up a sheet of A4 paper and ask the children what aspects of it you could measure. They will probably say that you could measure its length and width with a ruler. Tell them that you would like to find its mass and its volume (how much space it takes up). Let them discuss in pairs how they might do this, then ask for suggestions.

For mass, they may suggest using the weighing scales; explain that this is difficult with something so light, because the movement of the pointer will be too small to read. If the children don't come up with the idea, suggest that one way of overcoming this difficulty might be to weigh 50 or 100 sheets together and then divide the total mass by the number of sheets. Demonstrate this practically, using different children to help with the measuring and with the calculating.

With volume, the difficulty is that we rarely think of a piece of paper as having a thickness: we think of it as 2-D rather than 3-D. To convince the children that paper really does have a third dimension, show them a variety of different paper types with obviously different thicknesses. To calculate the thickness of one sheet, again use 50 or 100 sheets and divide the total thickness by the number of sheets. Now the volume can be calculated by multiplying the thickness by the length and the width.

Split the class in two for the next task, which they should carry out in pairs. Those who have found the reasoning hard to follow should repeat the work that you have led the class through. Children who you feel are capable of applying these ideas should be given the challenge of finding the mass and volume of a single playing card. They should report back their findings to the class.

DISCUSSION QUESTIONS
● *What do you know about this sheet of paper?*
● *Why is it difficult to measure its mass or its thickness?*
● *What is helpful about using 100 sheets of paper?*
● *Can you explain what you have found out about a playing card?*

ASSESSMENT
Do the children appreciate the difficulties involved in finding the mass and volume of a sheet of paper? Can they calculate the mass and volume of a sheet of paper (or a playing card) and explain how they found each?

EXTENSIONS
● The children can use the writing frame on resource page 128 to discuss their findings.
● They can compare the mass and volume of a range of different types of paper or card, or of different exercise books.

94

IMPERIAL MASS AND CAPACITY

†† Whole class, then individuals
⏱ 50 minutes

AIMS
To become familiar with imperial measures of mass and capacity. To convert between metric and imperial measures of mass and capacity.

WHAT YOU WILL NEED
Calculators, standard imperial masses, a pint bottle, a metric measuring cylinder, photocopiable pages 104 and 105, graph paper.

WHAT TO DO
NB You may prefer to discuss mass and capacity in separate sessions, using the corresponding photocopiable page in each case.

Talk to the children about imperial measures for capacity and mass. They may have some experience of pints, gallons, ounces, pounds and stones, but are unlikely to be familiar with the relationships between these units. Explain the relationships between imperial units: 8 pints = 1 gallon, 16 ounces = 1 pound, 14 pounds = 1 stone. These relationships can be used for a quick mental warm-up, asking questions such as:
- *How many pints are there in 4 gallons? [32]*
- *48 pints. How many gallons is that? [6]*
- *How many ounces are there in 5 pounds? [80]*
- *How many pounds are there in 12 stones? [168]*

Give practical demonstrations of the relationships between imperial and metric units, as follows:
● For capacity, fill a pint bottle with water. Do the children think that this will be more or less than a litre? Pour the water carefully into a litre measuring cylinder and measure the volume (around 0.57l or 570ml).
● For mass, use a pan balance. Put a 1lb mass on one side and add gram masses to the other until a balance is reached (around 450g).

Introduce either or both of the photocopiable sheets. All of the conversion information the children need for these tasks is given on the sheets. The conversions can be done with a calculator or, if you prefer, used for practice in multiplying decimals on paper. This is a good context for revising place value and decimal conventions; you may prefer the children to round their answers to the nearest whole number. In the second part of each sheet, the children will need to convert either from metric to imperial or from imperial to metric in order to make a comparison.

Review the children's answers together, and talk through any problems that they found difficult. The

answers to page 104 are: 1. a) 11lb, b) 19.8lb, c) 0.66lb, d) 26.4lb, e) 44lb, f) 7.14oz, g) 1.785oz, h) 10.71oz. 2. a) 18.9kg, b) 44.1kg, c) 3.6kg, d) 9kg, e) 20.25kg, f) 560g, g) 980g, h) 308g. 3. a) 3 kilograms, b) 15 ounces, c) 60 kilograms, d) 150 grams, e) 10 pounds. The answers to page 105 are: 1. a) 3.52 pints, b) 6.16 pints, c) 0.88 pints, d) 21.12 pints, e) 4.4 gallons, f) 5.5 gallons, g) 22 gallons, h) 17.6 gallons. 2. a) 3.42l, b) 6.84l, c) 11.4l, d) 57l, e) 31.92l, f) 68.4l, g) 91.2l, h) 282.72l. 3. a) 7 litres, b) 5 litres, c) 120 pints, d) 3 gallons, e) the same.

DISCUSSION QUESTIONS
● *What units of mass and capacity do you know?*
● *What things are measured in pints/gallons?*
● *What things are measured in stones/pounds?*
● *Which system of units is easier to understand, the metric or the imperial?*

ASSESSMENT
Can the children convert measures of mass and capacity between the metric and imperial systems? Do they appreciate the differences between the two systems of measures?

EXTENSIONS
● The children can create simple line graphs for converting between metric and imperial measures. It is useful to consider why these graphs will not give results as accurate as those of the numerical calculations; however, in many 'real life' situations, the level of accuracy provided by a conversion graph is quite acceptable.
● The children can use information books or CD-ROMs to find out about other systems of measurement that are used around the world.

WHICH UNITS?

†† *Whole class, then pairs*
🕐 *50 minutes*

AIM
To choose the most appropriate units for measuring different items.

WHAT YOU WILL NEED
Photocopiable page 106, pencils.

WHAT TO DO
Ask the children to think about what metric units they know. Ask them to discuss with a partner what each unit can be used to measure. They should identify not only what it measures (as in 'A kilogram measures mass'), but what the relative size of the unit is (for example, kilograms might be used for measuring a child's mass, whereas grams might be used for measuring the mass of a dormouse). After a few minutes, ask for examples. Spend some time discussing these.

Give each pair a copy of page 106. The pairs should agree on an appropriate unit for each item, bearing in mind both the type and the size of unit needed. Review their answers as a class.

DISCUSSION QUESTIONS
● *What units do you know?*
● *Which of those two units is on a larger/smaller scale?*
● *What would that unit be used to measure?*
● *Why is that a good/not a good unit to measure that item?*

ASSESSMENT
Do the children choose appropriate units to measure mass, length or capacity? Do they choose units on a sensible scale for particular items?

EXTENSION
The children can make up a nonsense story in which all the units are mixed up, then swap with a friend to find the mistakes. (There is a simple example of this in *Developing Shape, Space and Measures with 7–9 year olds*.)

HOW HEAVY IS WATER?

†† *Whole class, then groups*
🕐 *50 minutes*

AIMS
To distinguish between capacity and mass. To make and record accurate measurements, using a range of equipment.

WHAT YOU WILL NEED
Measuring cylinders, other containers, weighing scales, water, paper, pencils.

WHAT TO DO
Ask the children to discuss in groups how they could find the mass of 1 litre of water. They should make a note of all the equipment they need. To prompt their discussion, you could have a range of equipment (see above) prominently displayed.

Bring the class together and ask some groups for their ideas. Discuss with the class the pros and cons of the different suggestions. (One of the best ways is to weigh a small jar, pour 100ml of water into it, then weigh it again. The difference in mass will give you the mass of 100ml of water. This can be multiplied by 10 to find the mass of 1 litre.)

Let the groups try out their own methods. Remind them how to read the various scales and dials on the measuring equipment. Inaccuracies in these scales (or in the readings) may lead to slight variations in the final result; it may be useful to take an average of several groups' results in order to establish that 1 litre of water has a mass of 1kg. Relate this to the children's experience of fair tests in science.

MASS & CAPACITY

The children can write up their notes on this investigation, using the writing frame on page 128.

DISCUSSION QUESTIONS
● *How can you find the mass of water?*
● *What would be a good way to check that result?*

ASSESSMENT
Can the children explain how to find the mass of 1 litre of water? Can they make accurate measurements with a range of equipment? (Their results should be accurate enough for them to be confident that 1 litre of water does weigh exactly 1 kilogram.)

EXTENSIONS
● The children can investigate the mass of 1 litre of other liquids (such as milk, carbonated drinks, coffee and syrup), using the same techniques.
● This can be linked to science work on dissolving: how will dissolving sugar in warm water affect the mass of the water?

UNITS OF MASS AND CAPACITY

†† *Groups, then individuals*
🕐 *50 minutes*

AIM
To understand the relationships between different metric units of capacity (or volume of a liquid) and mass.

WHAT YOU WILL NEED
A collection of containers showing a variety of weights and fluid volumes (the labels will suffice – ask the children to collect these for a couple of weeks beforehand), paper, pencils, photocopiable page 107.

WHAT TO DO
Discuss the relationships between different units of mass, and between different units of capacity, as well as the standard abbreviations. For the activity, the children will need to know the following relationships:
● 1 litre (l) = 1000 millilitres (ml)
● 1 kilogram (kg) = 1000 grams (g)
Give a range of containers or labels to each group. They should sort these into ones that give a measurement of capacity (litres or millilitres) and ones that give a measurement of mass (kilograms or grams). They should then record these, and give each measurement an alternative scale. So, for example, a juice carton containing 350ml should be recorded as 350ml and 0.35l, while a label from a 2.5kg bag of potatoes should be recorded as 2.5kg and 2500g.

Other issues arising from an examination of different labels include the difference between 'net weight' and 'gross weight' ('net weight' is the mass of the contents only, 'gross weight' includes the mass of the container) and the possibility of other scales of units, such as centilitres or milligrams. Depending on the age and experience of the class, you may want to remove or deliberately include such items.

Photocopiable page 107 can be used for further individual consolidation of the relationship between units, either in the second half of this session or as a follow-up homework. The answers are:
A. 1) 2500g, 2) 600g, 3) 1605g, 4) 2085g, 5) 3004g, 6) 30g, 7) 4g, 8) 20 000g.
B. 4.5kg, 2) 0.7kg, 3) 2.307kg, 4) 1.064kg, 5) 2.009kg, 6) 0.025kg, 7) 0.009kg, 8) 0.999kg.
C. 1) 8650ml, 2) 900ml, 3) 3008ml, 4) 4050ml, 5) 6009ml, 6) 74ml, 7) 50ml, 8) 2ml.
D. 1.08l, 2) 0.8l, 3) 4.985l, 4) 4.07l, 5) 6.005l, 6) 0.06l, 7) 0.005l, 8) 1.001l.

MASS & CAPACITY

DISCUSSION QUESTIONS
● *What different (metric) units do we use for mass/capacity?*
● *What is the relationship between litres and millilitres/between kilograms and grams?*
● *What measure is given on this label?*

ASSESSMENT
Are the children familiar with metric measures for mass and capacity? Can they distinguish easily between the two types of measure? Can they convert between litres and millilitres? Can they convert between grams and kilograms?

VARIATION
You may want to split this activity into two separate sessions: one focusing on mass, the other on capacity.

EXTENSIONS
● The children can consider other metric relationships. When might it be appropriate to use any of the following?
1000 milligrams = 1 gram
1000 kilograms = 1 metric tonne
100 centilitres = 1 litre
1000 litres = 1 kilolitre
● Photocopiable page 108 requires the children to compare different units in order to solve problems. It can be used for individual follow-up work, or for whole-class discussion on this topic. The answers are: 1) 0.35kg, 2) ½ litre 3) 220g, 4) 450ml 5) 4½ kilograms 6) 11 litres.

BUY IN BULK?

†† *Pairs*
🕐 *50 minutes*

AIM
To solve problems involving mass, capacity and money.

WHAT YOU WILL NEED
Various different-sized packages of each of several foods (such as crisp packets, sweets, soft drinks, washing-up liquid, cereals) including labels or price tags, A3 paper, pens.

WHAT TO DO
Show the children some examples of different packages representing the same brand items, with their prices displayed. Ask them to discuss with a partner how they could decide which package is the best value.

Discuss the children's ideas. The solutions may vary according to the item – for example, the costs of a 25g bag and a 50g bag of the same brand of crisps can be compared directly (by doubling or halving), but the costs of a 500ml and a 325ml bottle of washing-up liquid will need a more sophisticated solution. The best solution is usually to calculate the value of each item in g or ml per penny (or pound), but the children should be encouraged to use a range of calculations that they feel comfortable with.

Give different sets of items to different pairs of children. They should write up their findings on a sheet of A3 paper, then present them to the class in a plenary session. These records can also be compiled to make a wall display or class book.

DISCUSSION QUESTIONS
● *Which size would you rather buy? Why?*
● *Would it be better to buy one of this size or two of that size?*
● *Who can explain why that bottle is better value?*

ASSESSMENT
Can the children find ways to compare prices? Can they explain their findings (for example, by recognizing that there is a trade-off between price and convenience)?

EXTENSIONS
● The class can visit a local supermarket to see how the prices are displayed in order to make comparisons easier. Someone from the supermarket may be able to talk to the class about this.
● The children can collect other product labels in order to make similar comparisons.

THE STRONGEST BAG

†† *Whole class, then groups*
🕐 *50 minutes*

AIMS
To weigh objects in a practical context. To plan and carry out a fair test.

WHAT YOU WILL NEED
A range of standard masses, particularly larger ones such as 1kg and 500g; other heavy objects (such as books or stones), weighing scales, a variety of paper and plastic bags, PE mats, A3 paper, pens.

WHAT TO DO
Show the children a range of paper and plastic bags. Ask them how they could find out which bag is the strongest – that is, which bag will hold the most mass. They are likely to realize that some will break when used to carry more than a certain load. Ask them to estimate how much mass any of the bags will carry before this happens.

Working in groups, the children should draw up a plan to find out which is the strongest bag. After a few minutes, ask them to report their plan back to the class. They can debate the pros and cons of the different ideas, then agree in their own groups as to how they will proceed.

Give each group four or five different bags and a PE mat, and make available a range of different masses, scales and heavy objects for them to choose from. Make sure that the groups work safely: the bags must be tested a few centimetres over a PE mat, not held up in the air or above the children's feet.

When they have completed their experiment, each group should draw a poster on A3 paper to present their findings to the class.

DISCUSSION QUESTIONS
● *How can you find out which bag is the strongest?*
● *Would that be a fair test? Why/why not?*
● *What mass did that bag hold before it broke?*
● *How can you check your measurements of mass?*

ASSESSMENT
Can the children suggest ways to solve the problem? Can they determine the mass that different bags will hold accurately? Can they explain their findings (by saying which materials are stronger)?

EXTENSION
The children can use the writing frame on resource page 128 to give an individual account of their work.

MASS & CAPACITY

MASS & CAPACITY

AN ABSORBING PROBLEM

†† *Whole class, then groups*
🕐 *50 minutes*

AIMS
To measure volumes of liquid accurately in a practical context. To plan and carry out a fair test.

WHAT YOU WILL NEED
A range of different absorbent materials (tissue paper, writing paper, kitchen cloths, towels, cotton cloth and so on) cut into various-sized pieces, measuring cylinders, water, newspaper or plastic sheeting (to cover working surfaces), A3 paper, pens. This activity might be carried out in the art or kitchen area.

WHAT TO DO
Start by pouring a small amount of water onto the floor. Show the children the different materials, and ask which would be the best one for mopping up the water. *Why will it be the best?* Use the suggested material to clear up the water. *What has happened to the water?* Check that they understand that the water has been 'absorbed' by the material.

Now ask the children how they could check which of the materials will absorb the most water, and how they could compare the 'absorbencies' of the materials. Ask them to make an accurate measurement, in millilitres, of how much water each material holds. They should discuss this in groups before sharing their ideas with the whole class. They will need to establish how to make the test fair – if they don't suggest it, you should prompt them to

consider using an identical-sized piece of each material.

The children should soak each of the materials (in turn) in water, then carefully squeeze out the material into a jug so that all of the water is collected. They should then measure the volume of water accurately. Alternatively, they could pour successive amounts of, say, 20ml onto the floor and then clear them up, until the material reaches saturation point.

After they have completed their experiment, each group should draw a poster on A3 paper to present their findings to the class.

DISCUSSION QUESTIONS
● *How can you find out which cloth holds the most water?*
● *What will you measure? What should be kept the same?*
● *Would that be a fair test? Why/Why not?*
● *How can you check your results?*
● *Can you do anything to make this experiment more accurate?*

ASSESSMENT
Can the children suggest ways to solve this problem? Can they record accurately the amount of water that different cloths will hold? Can they explain their findings by referring to the properties of materials?

VARIATION
The children can compare the absorbencies of different sponges in a similar experiment.

EXTENSION
The children can use the writing frame on resource page 128 to give an individual account of their work and findings.

WHICH HOLDS THE MOST?

†† *Whole class, then pairs*
🕐 *50 minutes*

AIM
To calculate the volumes of different boxes.

WHAT YOU WILL NEED
A range of different cuboid boxes (you could use those collected for the activity 'On the surface', page 83), rulers, pencils, paper, centimetre cubes.

WHAT TO DO
Show the children a pair of cuboid boxes that have different shapes – for example, a tin foil box and a chocolate box. These have obviously been designed for different purposes. Ask: *Why wouldn't you use each box for the other item?* Ask the children how they could compare the volumes of both boxes to find out which held the most. Ask them to consider filling each with either marbles or cubes. *Which would be best?* They should realize that small cubes (if they are carefully arranged) will be better, since all the space will be taken up; marbles always have spaces between them.

Show the children some centimetre cubes. *How could you work out how many of these will fit inside the box? What would you need to know to calculate this?* Explain that the volume of the box can be found by measuring the three dimensions of the box (its length, width and height) and multiplying them together. Discuss how finding a volume in this way can be compared to filling the box with centimetre cubes: in each case, the units of volume are centimetres cubed (cm³).

When the children are clear about this, give each pair some boxes. They should measure the length, width and height, then calculate the volume of each box.

It is useful to follow this activity by demonstrating the relationship between millilitres and centimetres cubed. Take a fairly small cuboid box. Invite a child to take its measurements and calculate its volume. Fill the box with sand, then pour the sand into a measuring cylinder: the measured volume of sand (in ml) should be equal to the volume of the box (in cm³). Any slight discrepancies are due to inaccuracies in the measurements.

DISCUSSION QUESTIONS
● *How can you compare the size of these boxes?*
● *What do we mean by the length, width and height of the boxes?*
● *Which box do you think has the largest volume? How could you check this?*

ASSESSMENT
Do the children understand how to find the volume of a cuboid? Can they measure the dimensions of a cuboid accurately and calculate its volume?

EXTENSIONS
● The children can make model cuboids from interlocking cubes. This is a useful way to confirm the idea that length multiplied by height multiplied by depth gives the volume of a cuboid. For example, a solid cuboid that is 3 × 4 × 5 cubes in size will contain 60 cubes in total.
● Give the children the dimensions of a range of cuboids and let them calculate the volume of each.
● The children can explore the relationship between the volume and the surface area of a cube or cuboid.
● This activity can be linked to science work on displacement. For example, the volume of a stone can be calculated by immersing it in a measuring cylinder half-full of water: since a cubic centimetre is equivalent to a millilitre, the increase in millilitres read on the cylinder's scale is equal to the volume of the stone. This technique allows the volume of irregular objects to be measured.

MASS & CAPACITY

MASS & CAPACITY

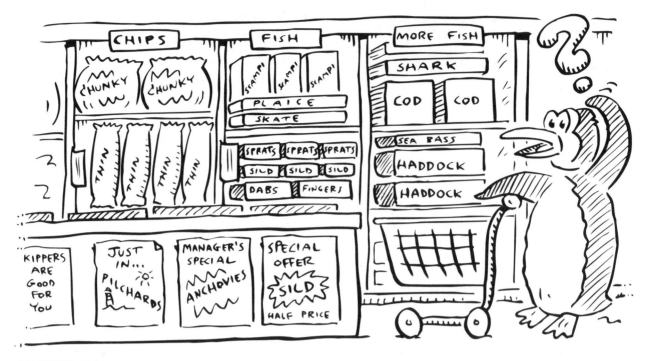

DOUBLE VOLUME?

†† Whole class, then pairs
⏱ 50 minutes

AIM
To consider (in a practical context) the effect of changing the dimensions of a box on its volume.

WHAT YOU WILL NEED
Various cuboid boxes (see 'On the surface', page 83), card, glue, scissors, adhesive tape.

WHAT TO DO
It is assumed that the children have already tried the activity 'Double up' (page 84), and so discovered that doubling the length and width of a rectangle actually multiplies its area by 4.

Show the children a cuboid box and say that their task is to design a box which will hold twice as much. They should discuss with a partner how to do this. Many children will suggest doubling each of the three dimensions – so remind them of their experience with doubling the size of a rectangle. In fact, to double the capacity of the box, only one dimension should be doubled. Put two similar boxes next to each other to demonstrate this, as it is quite difficult to visualize. If some children persist in believing that they must double each dimension, then leave them to do just that: they will find it easier to understand why this is wrong when they actually carry out the activity.

Give each pair a cuboid box and materials to make a new box with. Discuss the need to consider

the **net** of the original, and to draw a new net with certain measurements doubled. The new box should hold twice as much as the original. This can be tested approximately using Multilink cubes or similar, which will demonstrate that doubling all three dimensions is not the right method.

In a plenary session, the pairs can show their boxes and state the measurements of the original and the new box. Other children can check their calculations to confirm that one box has twice the volume of the other.

DISCUSSION QUESTIONS
● *Can you remember what happened with rectangles?*
● *What should we do to make the box hold twice as much?*
● *How can you make the new box? Which measurements will change?*

ASSESSMENT
Do the children understand the effect on volume of doubling any one of the dimensions? Do they understand the effect of doubling all three of the dimensions? Can they construct a new box accurately?

EXTENSIONS
● The children can use the writing frame on resource page 128 to discuss this activity.
● They can make a box in which each of the dimensions of the original is doubled (some may have chosen to do this anyway). *How many times larger than the original will this be?* [Eight times.]
● They can start with a large box (such as a cereal or washing powder box) and make a new box with half the volume.

CAKE RECIPES

†† *Whole class, then pairs or individuals*
🕐 *50 minutes*

AIMS
To reinforce familiarity with imperial measures of mass and capacity. To convert between metric and imperial measures of mass and capacity.

WHAT YOU WILL NEED
Calculators, photocopiable page 109, pencils, further recipes from cookery books, magazines or the Internet.

WHAT TO DO
Talk to the children about cookery recipes – what units are they likely to find in a recipe book?

Explain that the recipes on page 109 are from an old cookery book in which all the measurements are in the imperial system. Their task is to convert the measures into metric units, so that the school cook (who, of course, is very up-to-date) can make the food. Each conversion should be rounded to the nearest whole unit. Remind the children of the common unit conversions for mass and capacity

(these can be found on photocopiable pages 104 and 105, as well as in the 'Measures glossary' on pages 72–73).

The children can work on their own or with a partner – whichever you feel is more appropriate. You might like to have a further recipe ready for the class to work on together as a follow-up.

DISCUSSION QUESTIONS
● *What measures will you find in a cookery book?*
● *Are these imperial or metric units?*
● *Can you remember how to convert them?*

ASSESSMENT
Can the children convert measures of mass and capacity between the metric and imperial systems?

EXTENSIONS
● Tell the children that the school cook is off sick, and an older cook has taken over. Give them some recipes that use metric measures – can they convert these to imperial measures to help the other cook?
● Using the recipes on the worksheet, ask the children to find out how much of each ingredient would be needed to make half or twice as many cakes, or to make enough cakes for the whole school.

MASS & CAPACITY

CONVERTING MASS

| 1kg = 2.20lb | 100g = 3.57oz | 1st = 6.3kg | 1lb = 0.45kg | 1oz = 28g |

■ Use the information in the box above to make the following conversions:

1. Metric to imperial

    a) 5kg = _____

    b) 9kg = _____

    c) 0.3kg = _____

    d) 12kg = _____

    e) 20kg = _____

    f) 200g = _____

    g) 50g = _____

    h) 300g = _____

2. Imperial to metric

    a) 3 stones = _____

    b) 7 stones = _____

    c) 8 pounds = _____

    d) 20 pounds = _____

    e) 45 pounds = _____

    f) 20 ounces = _____

    g) 35 ounces = _____

    h) 11 ounces = _____

3. Which is the heavier in each case?

    a) 3 kilograms or 6 pounds? _____

    b) 15 ounces or 400 grams? _____

    c) 9 stone or 60 kilograms? _____

    d) 150 grams or 5 ounces? _____

    e) 10 pounds or 4 kilograms? _____

 Which system do you prefer to use, the imperial or the metric?
Do your friends and family agree?

| SEE 'IMPERIAL MASS AND CAPACITY', PAGE 95. |

NAME _____ DATE _____

CONVERTING CAPACITY

1 litre = 1.76 pints	1 pint = 0.57 litres
1 litre = 0.22 gallons	1 gallon = 4.56 litres

■ Use the information in the box above to make the following conversions:

1. Metric to imperial

   a) 2l = _____ pints

   b) 3.5l = _____ pints

   c) 0.5l = _____ pints

   d) 12l = _____ pints

   e) 20l = _____ gallons

   f) 25l = _____ gallons

   g) 100l = _____ gallons

   h) 80l = _____ gallons

2. Imperial to metric

   a) 6 pints = _____

   b) 12 pints = _____

   c) 20 pints = _____

   d) 100 pints = _____

   e) 7 gallons = _____

   f) 15 gallons = _____

   g) 20 gallons = _____

   h) 62 gallons = _____

3. Which is the greater amount in each case?

   a) 10 pints or 7 litres? _____

   b) 5 litres or 1 gallon? _____

   c) 120 pints or 67 litres? _____

   d) 13 litres or 3 gallons? _____

   e) 200 litres or 44 gallons? _____

 On the back of the sheet write some instructions for a friend, explaining how to solve these problems.

SEE 'IMPERIAL MASS AND VOLUME', PAGE 95.

CHOOSE THE RIGHT UNITS

■ Write down the most sensible units to use for each of the following measurements.

Length of a football field	Weight of an exercise book
Distance to the moon	Amount of water in a bath
Height of a person	Amount of medicine in a bottle
Width of a paper clip	Weight of an elephant
Weight of a pencil sharpener	Amount of juice in a glass

 Does your friend agree?

 Could some questions have more than one answer?

SEE 'WHICH UNITS?', PAGE 96.

NAME   DATE

METRIC MASS AND CAPACITY

Remember: 1kg = 1000g   1l = 1000ml

A. Rewrite these masses in grams.

1) 2.5kg = _____

2) 0.6kg = _____

3) 1.605kg = _____

4) 2kg 85g = _____

5) 3kg 4g = _____

6) 0.03kg = _____

7) 0.004kg = _____

8) 20kg = _____

C. Rewrite these capacities in millilitres.

1) 8.65l = _____

2) 0.9l = _____

3) 3.008l = _____

4) 4l 50ml = _____

5) 6l 9ml = _____

6) 0.074l = _____

7) 0.05l = _____

8) 0.002l = _____

B. Rewrite these masses in kilograms.

1) 4500g = _____

2) 700g = _____

3) 2307g = _____

4) 1kg 64g = _____

5) 2kg 9g = _____

6) 25g = _____

7) 9g = _____

8) 999g = _____

D. Rewrite these capacities in litres.

1) 1080ml = _____

2) 800ml = _____

3) 4985ml = _____

4) 4l 70ml = _____

5) 6l 5ml = _____

6) 60ml = _____

7) 5ml = _____

8) 1001ml = _____

 How would you help a friend who found this exercise confusing?

SEE 'UNITS OF MASS AND CAPACITY', PAGE 97.

**DEVELOPING SHAPE,
SPACE & MEASURES**

NAME

DATE

WHICH IS GREATER?

■ Solve these problems. Explain how you worked them out.

Which is greater?

1. A loaf of bread that weighs 300g or one that weighs 0.35kg? 	2. A beaker that holds $\frac{1}{2}$ litre or one that holds 400ml?
3. A bunch of bananas weighing 0.022kg or one that weighs 220g? 	4. A bottle with 450ml of lemonade or one with 0.045l? 
5. A baby weighing $4\frac{1}{2}$ kilograms or one weighing 4050 grams? 	6. A bucket with a capacity of 11 litres or one with a capacity of 10 100ml?

 Make up some more puzzles like these for your friends

SEE 'UNITS OF MASS AND CAPACITY', PAGE 97.

MASS & CAPACITY

NAME _____   DATE _____

MAKING CAKES

■ Convert the amounts in these recipes to metric measurements.

SCONES
– makes 16 scones

8oz self-raising flour _____

1oz caster sugar _____

2oz butter _____

$\frac{1}{4}$ pint milk _____

SHORTBREAD
– makes 8 pieces

5oz plain flour _____

1oz ground rice _____

4oz butter _____

2oz caster sugar _____

CHOCOLATE BUNS
– makes 18 buns

$2\frac{1}{2}$ oz plain flour _____

2oz butter _____

$\frac{1}{4}$ pint cold water _____

$\frac{1}{2}$ pint double cream _____

9oz plain chocolate _____

SPONGE CAKES
– makes 20 cakes

8oz soft margarine _____

8oz caster sugar _____

3oz butter _____

10oz self-raising flour _____

6oz icing sugar _____

 Check your answers with a friend – do you agree?

 Now find a cookery book. What units does it use?

SEE 'CAKE RECIPES', PAGE 103.

DEVELOPING SHAPE,
SPACE & MEASURES

KEY IDEAS

- Consolidating the vocabulary of time.
- Estimating the length of events.
- Using 24-hour clock times.
- Reading and understanding timetables.
- Solving problems involving the length of different events.

Teachers' experience suggests that the teaching and learning of time concepts does not occur in a simple way. Children's experiences outside school may play a major part in this, with some children being helped with the basics of learning to tell the time and some households being accurate and punctual in their use of time (for example, always having Sunday lunch at 1pm). Many children now have digital watches, and digital displays of time are evident everywhere – for example, on video recorders and timetables.

One assumption made in these activities is that the children are already familiar with the conventions of reading an analogue clock face – for children not confident with this, some useful activities can be found in the companion Scholastic book *Developing Shape, Space and Measures with 7–9 year olds* (also by Jon Kurta).

A key aspect of work in this area is to link whatever representation of time is being used (clock faces or digital displays) to real events. The school daily timetable, a local bus timetable and a television schedule are all useful for this.

Children also need to have opportunities to consolidate their understanding of the different units of time and the scales that they represent. As with the imperial units of length or mass, a range of different numerical relationships exist between different time units. The children need to appreciate which units are appropriate for different contexts, how to measure each, and how different the scales of different units are without having a simple metric structure to rely on.

BY THE END OF Y5/P6, MOST CHILDREN SHOULD BE ABLE TO:

- use units of time
- estimate periods of time and suggest suitable units (including for events outside their own experience)
- read the time on a 24-hour digital clock and use 24-hour clock notation
- use simple timetables and work out journey times.

BY THE END OF Y6/P7, MOST CHILDREN SHOULD BE ABLE TO:

- appreciate the idea of time zones around the world
- use 'am' and 'pm' and the notation 9:53
- read measures of time in seconds from a conventional stop-watch and more accurately from a digital stop-watch
- read more complex timetables.

SOME COMMON MISCONCEPTIONS AND STRATEGIES FOR CORRECTING THEM

24-HOUR CLOCK CONVENTIONS

When converting from 24-hour to 12-hour clock times, children sometimes just take off 10 – for example, stating that 19:30 is 9:30pm. They need to be reminded that 'pm' times contain 12 hours (once around the clock) more than 'am' times. Talk through the transition from 12:00 (noon) to 13:00 (1:00pm) and 14:00 (2:00pm).

24-hour clock times always have four digits. When they are converting from 12-hour to 24-hour times, children often need to be reminded to put a zero before a single-digit hour: 7:25am is 07:25.

Similarly, children often need to be reminded to add 12 when converting a 'pm' 12-hour time to a 24-hour time: 6:45pm is 18:45.

Note that for writing 24-hour clock times, several conventions may be used: 19.30, 19:30 or 1930. In this book, 19:30 is used consistently to avoid possible confusion with other measures.

CALCULATIONS WITH TIME

Because time is not measured in metric units, normal vertical calculation methods are often misleading where amounts of time are involved. It is far better to help the children develop an intuitive feel for these, based on mental strategies. For example:

Train leaves London at 08:50 and arrives at Glasgow at 13:15. How long does it take?
Encourage the children to use a counting on method: 10 minutes (to 09:00) + 4 hours (to 13:00) + 15 minutes (to 13:15) = 4 hours 25 minutes. The vertical subtraction method would require the children to 'borrow' an hour as 60 minutes (since time is not a metric measure):

13:15	→	12.75		
– 08:50		– 08.50	→	04.25

Here, the conventional written method is fraught with difficulties. As well as being easier to check, the counting on method is a more accurate reflection of the children's experience of time.

TIME

UNITS OF TIME

†† *Whole class, then individuals*
🕐 *50 minutes*

AIMS
To revise units of time. To understand the relationship between different units of time.

WHAT YOU WILL NEED
A table showing the relationships between different units of time (see 'Measures glossary', page 72–73), photocopiable page 116, pencils, calculators.

WHAT TO DO
Start by asking the children to list, with a partner, all the different units of time that they know. After a few minutes, ask the children to feed these back; when a complete list has been compiled, ask the children to list the units in order of scale. The list should range from 'second' to 'millennium'. When the order has been agreed, ask the children to think about the relationships between different units.

Say that the units of time are more like imperial units than metric ones, in that there is a whole range of different relationships to remember between the different units, rather than a simple scale of tens.

Now ask the children how they would answer questions such as:
● *How many seconds are there in an hour?*
● *How many minutes are there in a day?*
● *How many days are there in a decade?*
Note that each of these problems requires a 'double calculation', using an intermediate unit. For example, to find out how many seconds there are in an hour, you need to think first about the number of seconds in a minute, then about the number of minutes in an hour.

After the children have worked through a few examples, introduce page 116. Here, the children need to use the relationships between units of time in order to establish which of two periods is greater. The final question at the bottom of the sheet is an interesting one for follow-up discussion. The answers are: 1) 3 hours, 2) 6000 minutes, 3) 7 weeks, 4) 4 decades, 5) 21 centuries, 6) a fortnight, 7) a decade, 8) 3 days.

DISCUSSION QUESTIONS
● *Which time units relate to long periods? Which are very short?*
● *Who can tell me how many seconds are in an hour? How many minutes are in a week?*
● *Which time units do we use most? Which are the important relationships to remember?*

ASSESSMENT

Do the children know the different time units and the relationships between them? Can they explain how to compare time periods given in different units?

EXTENSION

Develop a group discussion on the theme: *What would be the effect if the units for time were metricated? What would a day be like? What would a year be like?* The different units must all now have relationships of 10, 100 or 1000.

24 HOURS

†† Whole class, then individuals
🕐 50 minutes

AIMS

To consider the duration of events. To calculate with time.

WHAT YOU WILL NEED

Paper, pencils.

WHAT TO DO

Ask the children to imagine that they have 24 hours to spend doing their favourite things. How will they spend their time? List some of their favourite things, and ask them how long they will spend on each. Take a few responses, then ask the children to consider how they would spend the 24 hours. Let them choose their own way to present this – for example, a written account, a list, a graph or a formal timetable. However, insist that during the 24 hours, they must spend realistic amounts of time on a variety of activities: no 24-hour television or pizza marathons are allowed (though sleep can be minimized). The time period covered must be exactly 24 hours.

The children should work individually, though they might collaborate on finding ways to present their timetable.

DISCUSSION QUESTIONS

● *What do you think you could do in 24 hours?*
● *How long would you watch television/play football/ visit friends?*
● *How can you check that you have filled exactly 24 hours?*

ASSESSMENT

Can the children compile a 24-hour timetable that shows clearly how the time will be spent? Can they calculate times as necessary?

VARIATIONS

● The children can work out a timetable for 12 hours, or for a week.
● Ask the children to base their account on real events in a typical school day, a day at the weekend or a day in the school holiday. In this case, the 24 hours ought to include a reasonable period of sleep!

TIME

50

TIMESHEET

†† *Whole class, then individuals*
🕐 *50 minutes*

AIM
To write times using the 24-hour clock system.

WHAT YOU WILL NEED
A variety of different clocks and watches (or pictures of them), paper, pencils, photocopiable page 117.

WHAT TO DO
With the class, examine a variety of different clocks and watches. Discuss the different ways in which they represent the time, using different symbols and arrangements.

Explain the conventions for writing times using the 24-hour clock (see page 111 for notes on common misconceptions and sources of confusion). Refer to various events in the school day and ask the children to write these times, using both the 12-hour and the 24-hour clock systems. For example:
● school assembly at a quarter past nine, written as either 9:15am or 09:15
● 'home time' at half past three, written as either 3:30pm or 15:30.

Give each child a copy of photocopiable page 117. The children have to write the time for a variety of events in words and then in numbers, using both systems. Review the answers as a class.

DISCUSSION QUESTIONS
● *What different ways are there of writing the time?*
● *Which system do you find easiest to use?*
● *What are the advantages/disadvantages of each system?*

ASSESSMENT
Do the children understand the conventions for writing 12-hour and 24-hour clock times? Can they convert times between the two systems?

EXTENSIONS
● Photocopiable page 118 provides practice in reading and ordering times presented using different numerical systems, and in writing the same times in words.
● Find a TV schedule presented using 12-hour clock times. Ask the children to rewrite the schedule using 24-hour clock times.
● Find a bus or train timetable presented using 24-hour clock times. Ask the children to rewrite it using 12-hour clock times. (You could use the timetable provided for 'Timetable' opposite.)

TIMETABLE

†† Whole class, then pairs
🕐 50 minutes

AIMS
To read timetables. To solve problems involving time.

WHAT YOU WILL NEED
Local bus or train timetables, local maps, photocopiable page 119.

WHAT TO DO
The details of this activity will depend on the local transport information that you have. Using the questions on page 119 as models, compile some questions for the children to answer about local travel timetables. You may prefer to cut out a sample of a real timetable and copy it onto an OHT to discuss with the class (some bus and train timetables are difficult or confusing to read because of the small print used). Establish with the class that the times are in 24-hour format, and remind them of previous work in that area.

The types of question to ask include the following:
● Reading the timetable – *at what time does the bus or train arrive at a certain place?*
● Duration – *how long will the bus or train take to get from A to B? Is this time always the same?*

● Arrival – *will the bus or train get to a particular place by a particular time? [Converting 24-hour times to 12-hour times.]*
After discussing these questions in relation to your local timetables, give each pair a copy of page 119 and let the children answer the questions on it collaboratively. Review the work as a class. The answers are: 1) 11:20, 2) 13:45, 3) 16:00, 4) 10:30, 5) 12:55, 6) 16:00, 7) 2 hours 5 minutes, 8) 2 hours 45 minutes, 9) 1 hour 25 minutes, 10) 3 hours 15 minutes. Give the pairs opportunities to pose further questions about the timetable on page 119 for the class to answer.

DISCUSSION QUESTIONS
● *Who can explain how the timetable works?*
● *What time will that train get to its last station?*
● *Which is the fastest/slowest train between these two stations?*

ASSESSMENT
Do the children understand how to read a timetable? Can they do calculations based on the timetable?

EXTENSIONS
● Ask the children to rewrite a timetable using 12-hour clock times.
● Ask them to extend the timetable on page 119 for further trains, or for trains making the return journey from Capital Centre to Newtown Central.

NAME   **DATE**

THE LONGER TIME

■ In each case, decide which is the longer period of time and explain why.

1) 10 000 seconds or 3 hours?	2) 4 days or 6 000 minutes?
3) 7 weeks or 1000 hours?	4) 4 decades or 14 000 days?
5) 21 centuries or 2 millennia?	6) 20 000 minutes or a fortnight?
7) A decade or 10 000 hours?	8) 3 days or 250 000 seconds?

 Do you think your teacher has lived more than a million hours? Discuss this with your friends.

SEE 'UNITS OF TIME', PAGE 112.

NAME

DATE

A WHOLE DAY

■ Complete the table, writing the times that you do these things – first in words, then in the 12-hour (am and pm) and 24-hour clock systems.

	Time in words	12-hour time	24-hour time
Get up			
Leave for school			
School starts			
Morning break			
Lunchtime			
End of school			
Teatime			
Do my homework			
Watch TV			
Go to bed			

 Draw up a similar table for a typical day at the weekend.

SEE 'TIMESHEET', PAGE 114.

**DEVELOPING SHAPE,
SPACE & MEASURES**

TIME

PUT THE TIMES IN ORDER

■ Write the following twelve times in the right places on the 24-hour ladder.
■ Next to each time, write the same time in words. For example:
3:20am – twenty past three in the morning.

3:20am	5:30pm	11:45am	8:40pm	1:15pm	7:10am
09:30	12:30	18:45	01:20	02:15	15:35

Midnight

Noon

Midnight

 Compare your time ladder with a friend's. Do you agree?

SEE 'TIMESHEET', PAGE 114.

NAME _____ DATE _____

TRAIN JOURNEY

■ Use the train timetable to answer the questions below.

	Train A	Train B	Train C	Train D
Newtown Central	08:30	10:30	11:30	14:00
Parkview Lane	09:15	11:00	12:15	15:10
Shopping City	09:55	…	12:55	16:00
Seaside View	10:45	…	13:45	16:45
Woodbridge	11:20	12:20	14:20	17:30
Capital Centre	12:05	13:15	15:05	18:25

1. At what time does Train A arrive at Woodbridge?
2. At what time does Train C arrive at Seaside View?
3. At what time does Train D arrive at Shopping City?
4. I need to be at Parkview Lane for 11:15am. When should I catch the train at Newtown Central?
5. I need to be at Woodbridge for 3pm. When should I catch the train at Shopping City?
6. I need to be at Capital Centre for 6:30pm. When should I catch the train at Shopping City?
7. How long does Train A take to go from Parkview Lane to Woodbridge?
8. How long does Train B take to go from Newtown Central to Capital Centre?
9. How long does Train C take to go from Shopping City to Woodbridge?
10. How long does Train D take to go from Parkview Lane to Capital Centre?

 On the back of the sheet, write four more questions about these trains. Check that you know the answers, then give them to your friend to try.

SEE 'TIMETABLE', PAGE 115.

TRIANGLES

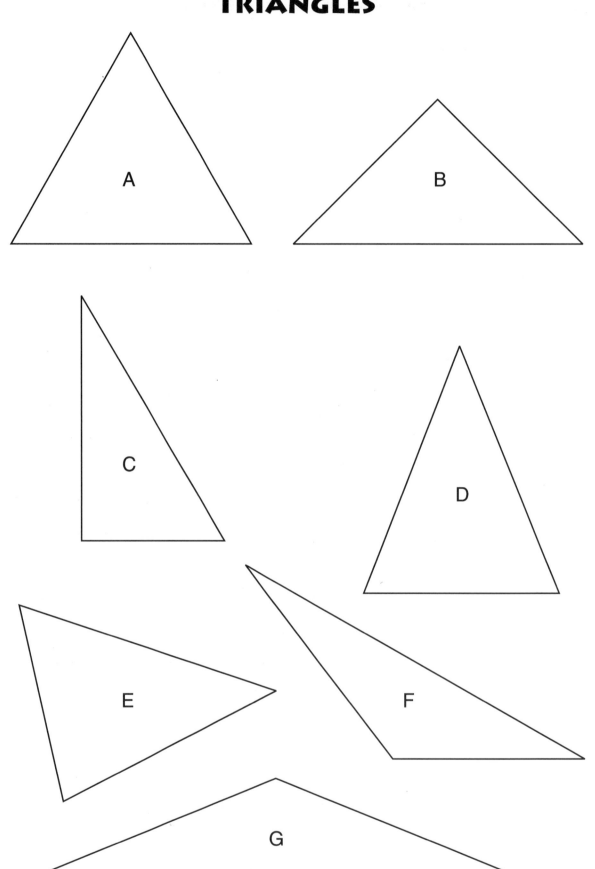

PLEASE REFER TO PAGES 13, 31, 59, 63.

**DEVELOPING SHAPE,
SPACE & MEASURES**

QUADRILATERALS

A

B

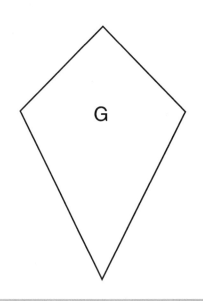

C

D

E

F

G

H

PENTOMINOES

1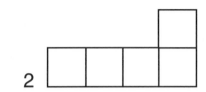

2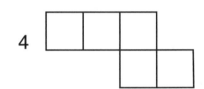

3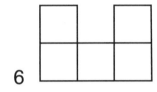

4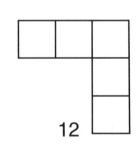

5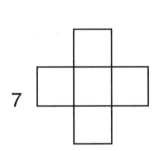

6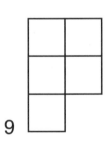

7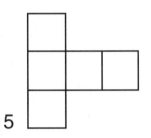

8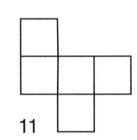

9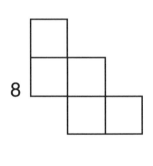

10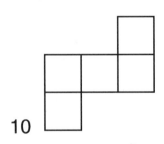

11

12

PLEASE REFER TO PAGES 16, 18, 32, 33, 82.

**DEVELOPING SHAPE,
SPACE & MEASURES**

HEXOMINOES

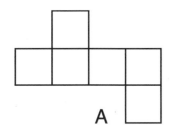

A

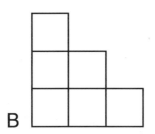

B

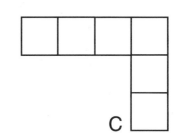

C

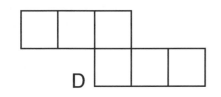

D

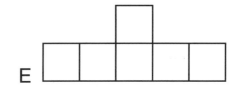

E

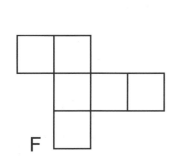

F

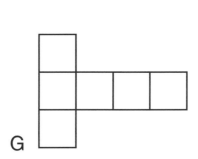

G

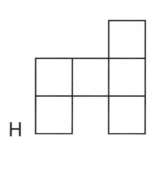

H

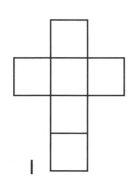

I

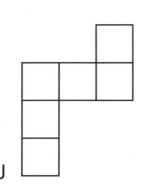

J

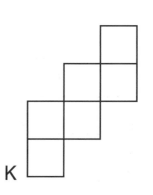

K

PLEASE REFER TO PAGES 16, 18, 32, 33, 82.

**DEVELOPING SHAPE,
SPACE & MEASURES**

NAME   DATE

'FOLLOW THE SHAPE' GAMECARDS

I am a circle.	I am an irregular pentagon.	I am a regular pentagon.
I have four equal sides and four right angles.	*I am a quadrilateral. My opposite sides are parallel. I have no right angles.*	*I have three sides. Each side is a different length.*
I am a square.	I am a parallelogram.	I am a scalene triangle.
I have three sides and one line of symmetry.	*I have six equal sides and six equal angles.*	*I have one straight line and one that is curved.*
I am an isosceles triangle.	I am a regular hexagon.	I am a semicircle.
I have eight equal sides.	*I have three sides and three lines of symmetry. Each of my angles is 60°.*	*I have six sides. All of my angles are different.*
I am a regular octagon.	I am an equilateral triangle.	I am an irregular hexagon.
I am a quadrilateral with four right angles and two lines of symmetry.	*I am a quadrilateral. My diagonals cross at right angles. I have two pairs of adjacent sides that are of equal length.*	*I am a quadrilateral. My sides are all of equal length. I have no right angles.*
I am a rectangle.	I am a kite.	I am a rhombus.
I have five sides and no lines of symmetry.	*I have five sides and five lines of symmetry.*	*I have no corners. A full turn around my centre is 360°.*

PLEASE REFER TO PAGE 20.

GEOBOARDS

PLEASE REFER TO PAGE 30.

125

6 × 6 GRIDS

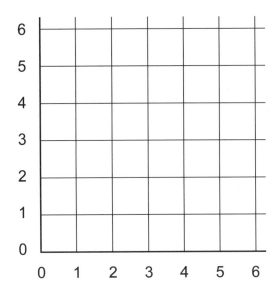

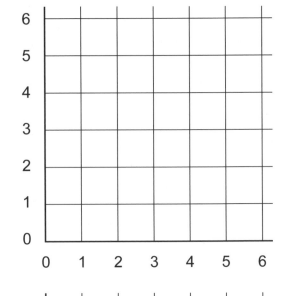

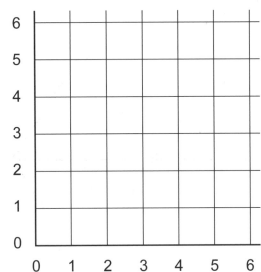

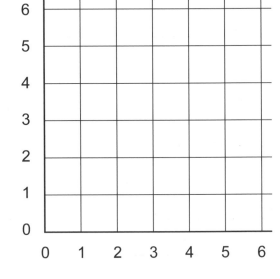

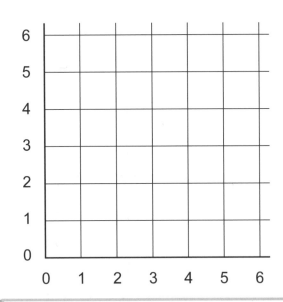

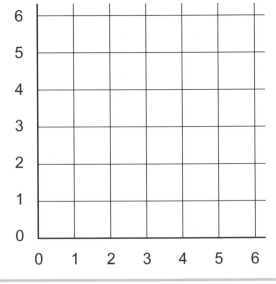

PLEASE REFER TO PAGE 49.

**DEVELOPING SHAPE,
SPACE & MEASURES**

LOGO PATTERNS

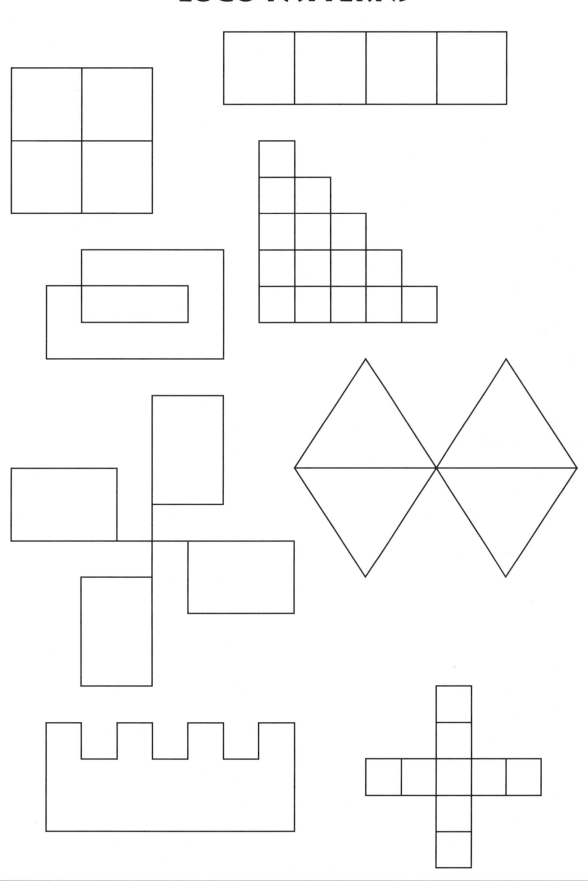

PLEASE REFER TO PAGE 48.

**DEVELOPING SHAPE,
SPACE & MEASURES**

WRITING FRAME

Today we investigated...

We used...

We found out...

I have learnt that...

Some important maths words I used: